The Waking Dream of T. E. Lawrence

The Waking Dream of T. E. Lawrence

Essays on His Life, Literature, and Legacy

Edited by Charles M. Stang

palgrave

THE WAKING DREAM OF T. E. LAWRENCE

First published 2002 by PALGRAVE™
175 Fifth Avenue, New York, N.Y. 10010 and
Houndmills, Basingstoke, Hampshire RG21 6XS.
Companies and representatives throughout the world

PALGRAVE is the new global publishing imprint of St. Martin's Press LLC Scholarly and Reference Division and Palgrave Publishers Ltd (formerly Macmillan Press Ltd).

ISBN 0–312–23757-X

Library of Congress Cataloging-in-Publication Data
The waking dream of T. E. Lawrence : essays on his life, literature and legacy / edited by Charles M. Stang.
p. cm.
Includes bibliographical references and index.
ISBN 0–312–23757-X
1. Lawrence, T. E. (Thomas Edward), 1888–1935. 2. Lawrence, T. E. (Thomas Edward), 1888–1935—Knowledge and learning. 3. History—Philosophy. 4. Scholars—Great Britain. 6. Orientalists—Great Britain. 7. Military art and science—Middle East—History—20th century. 8. Middle East—History, Military. I. Stang, Charles M.

D568.4.L45 W35 2002
940.4'15'092—dc21 2001053135

A catalogue record for this book is available from the British Library.

Design by Letra Libre, Inc.

First edition: March 2002
10 9 8 7 6 5 4 3 2 1

Printed in the United States of America.

Contents

All men dream: but not equally. Those who dream by night in the dusty recesses of their minds wake in the day to find that it was vanity: but the dreamers of the day are dangerous men, for they may act their dream with open eyes, to make it possible. This I did.

—T. E. Lawrence

ου φροντις

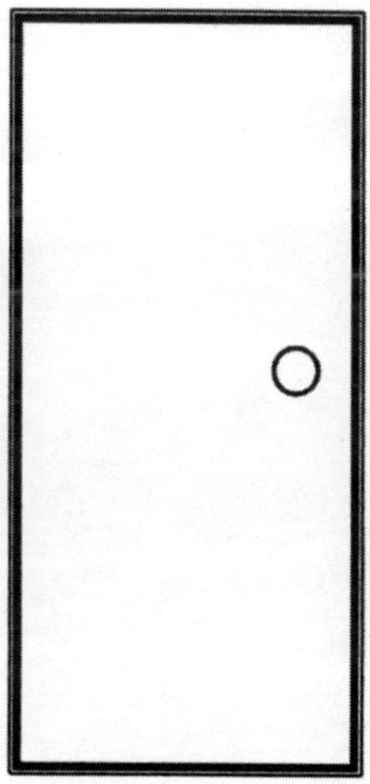

All these are scraps. The real framework, the place which his spirit will never cease to haunt, is Clouds Hill, and the gay motto over his doorway is the one beneath which I see him at rest.

—E. M. Forster

Contributors

MAREN COHN received her B.A. from Stanford University and her M.A. and Ph.D. from the Committee on Social Thought at the University of Chicago. She has published work and lectured on Nietzsche, Joyce, Lawrence, and the *Odyssey,* and she is currently writing an essay on Herman Melville's *Billy Budd.* She is a tutor at St. John's College in Santa Fe, New Mexico.

GEORGE W. GAWRYCH received his Ph.D. from the University of Michigan in 1980. Since July 1984, he has held the position of Historian at the U.S. Army Command and General Staff College where he specializes in conducting research and teaching courses on modern Middle Eastern and Balkan history. His book *The Albatross of Decisive Victory: War and Policy Between Egypt and Israel in the 1967 and 1973 Arab-Israeli Wars* was published by Greenwood Press in March 2000. Other publications include two monographs and numerous articles on Middle Eastern and Balkan history. In November 1989, he received the Turkish Studies Association's biennial prize for the best published article in Turkish studies. Currently, he is studying the U.S. Army's peace support operations in Bosnia, a project funded by the United States Institute for Peace.

JAMES L. GELVIN is Associate Professor of history at the University of California, Los Angeles. He is the author of *Divided Loyalties: Nationalism and Mass Politics in Syria at the Close of Empire* (Berkeley: University of California Press, 1998). A specialist in the cultural and social history of the nineteenth- and twentieth-century Middle East, Professor Gelvin has published articles in the *International Journal of Middle East Studies,* the *Journal of Interdisciplinary History, Nations and Nationalism, World Affairs,* and *International History Review,* as well as chapters in several edited volumes.

JONATHAN HART is a Professor at the University of Alberta and the author of *Theater and World* (1992), *Northrop Frye* (1994), *Breath and Dust* (2000), and *Representing the New World* (2001). His specialty is in literature and history, and he has held visiting appointments at Toronto, Harvard, Cambridge, Princeton, and elsewhere.

STEPHANIE NELSON received her B.A. from St. John's College and her M.A. and Ph.D. from the Committee on Social Thought at the University of Chicago. She currently teaches courses in Classics and in the Core Curriculum at Boston University.

CHARLES M. STANG received his A.B. from Harvard University and his MDiv. from the University of Chicago. He was the 1997 - 98 Annenberg Fellow at Eton College, where he taught American Studies, and he was a Contributing Writer for *The Boston Book Review.* He currently lives in Cairo.

THEOHARIS CONSTANTINE THEOHARIS lectures widely in Europe and the United States. He is the author of *Joyce's Ulysses: An Anatomy of the Soul* (1988) and *Ibsen's Drama: Right Action and Tragic Joy* (1996). His translation of the complete poems of Constantine P. Cavafy, *Before Time Could Change Them,* was published in April 2001 by Harcourt Inc. He teaches at Harvard University.

Acknowledgments

It is a right and good and joyful thing to give thanks and praise to those who have helped this small book come into being. I am grateful for the patient efforts of my editors at Palgrave, Kristi Long and Michael Flamini. Many others at Palgrave have offered a hand at various stages along the way and deserve mention, including Meg Weaver, Roee Raz, Enid Stubin, and Sabahat Chaudhary. Particular thanks is due to Theoharis C. Theoharis for his early support of and participation in this project. To David Tracy I owe the pleasure of hours of conversation on Lawrence and the introduction to a seventh contributor. To Chad Butler, John Seal, and Paul-Jon Benson I owe thanks for their friendship in Chicago, where this book was conceived and birthed. I wish also to thank my family, especially my parents Paul and Theodora, for their patience and enthusiasm. Finally, I wish to thank Sarabinh Levy-Brightman for her loving support and keen editorial eye. These pages of mine are, like my life, much improved by their contact with her.

Selections from the writings of T. E. Lawrence are published with permission of the Seven Pillars of Wisdom Trust.

Selections from *Seven Pillars of Wisdom* by T. E. Lawrence are published with permission of Doubleday, a division of Random House, Inc.

Selections from *The Odyssey of Homer* translated by T. E. Lawrence are published with permission of Oxford University Press, Inc.

Selections from the letters of T. E. Lawrence to F. N. Doubleday and Sir John Maxwell are published with permission of the Princeton University Library.

Selections from *Untimely Meditations* by Friedrich Nietzsche are reprinted with the permission of Cambridge University Press.

Introduction

The Many Ways of T. E. Lawrence

Charles M. Stang

In December 1927, on the strength of his singular work *Seven Pillars of Wisdom,* T. E. Lawrence was invited by Bruce Rogers, a talented American typographer, to undertake a new translation of Homer's *Odyssey*—by Lawrence's reckoning the twenty-eighth such translation into English. Nearly ten years since World War I and the Arab Revolt that won for him the epithet "of Arabia," Lawrence was at that time in the ranks of the Royal Air Force, fleeing from the prison of his own legend in a remote air station in Karachi, India. He therefore accepted Rogers's invitation on the condition that a "virgin" name, "Shaw," be attached to the translation.[1]

In the first line of the *Odyssey,* Lawrence would encounter a term that might as equally describe himself as Odysseus: πολυτροπος. Lawrence renders it as "various-minded man," but I prefer, for both Odysseus and Lawrence, "man of many ways."[2] The many ways of Lawrence are often overshadowed by his monolithic myth. One scholar reminds us that Lawrence was "archaeologist, intelligence agent, guerrilla leader, military theorist, diplomat, writer, translator, book designer, mechanic and guilt-scarred flagellant, lover and hater of publicity, and boyish leg puller."[3] Informing this catalogue is a set of equally divergent yearnings: disclosure or concealment, fame or anonymity, loyalty or betrayal, society or isolation, integrity or violation—these were, in Emerson's words, the poles between which his being swung. And swing he did, but when the tension between these poles grew too great, he was left in ruins. In such a mood, he could write to friends that his "two selves, you see, are mutually destructive. So I fall between them into the nihilism which cannot find, in being, even a false God in which to believe."[4]

And yet, as was true for Odysseus, these many ways were the source of Lawrence's remarkable achievements: military, political, and literary. This collection of essays attempts to take stock of these polytropic achievements, from Lawrence's articulating an innovative method of irregular warfare to his writing a history on the model of Herodotus to his rendering Odysseus into English anew. Lawrence himself was often prevented from appreciating his own achievements by what André Malraux calls "The Demon of the Absolute."[5] Lawrence writes: "One of the sorest things in life is to come to realize that one is just not good enough. Better perhaps than some, than many, almost—but I do not care for relatives, for matching myself against my kind. There is an ideal standard somewhere and only that matters: and I cannot find it. Hence this aimlessness."[6] If we wish to understand his achievements, we must balance his disappointment with the testimony of those who knew Lawrence well—testimony as honest as from his Syrian foreman (Sheikh Hamoudi), "Tell them, tell them in England what I say. Of manhood the man, in freedom free; a mind without equal; I can see no flaw in him,"[7] or as monumental as from a British prime minister (Winston Churchill): "I deem him one of the greatest beings alive in our time. I do not see his likeness elsewhere. I fear whatever our need we shall never see his like again."[8] The truth of Lawrence, if indeed there is any such thing, hangs between.

In response to critics' efforts to dispel the clouds of mystery surrounding Lawrence, Miss Fareedah el Akle, Lawrence's Arabic teacher from before the war, wrote the following: "Lawrence seems to me to be like an oyster which has, through pain and suffering all through life, developed into a pearl which the world is trying to evaluate, taking it to pieces layer by layer, without realizing the true value of the whole."[9] To strike a balance between part and whole is the task of any study, but it is all the more crucial for Lawrence, this man of many ways. In this case, a narrow focus obscures the integrity of Lawrence's many ways, while the wide view often neglects the depth of his genius. And in her description of Lawrence as a pearl, Miss Akle pleads for tenderness and calls our attention to the suffering and the fragility of this rare subject, who was, after all, not a myth, but a man. Here, then, are just a few of the layers of the Lawrence pearl, some of his many ways.

Student of History

Lawrence could hardly act in the world without reference to history. The present was for him the forge in which the future could be fashioned according to the model of the past. He was not a reactionary, but a romantic. Present everywhere is his conviction that time moves in grand patterns, that the key for divining what could be done lay in what had been done. What was being wrought in the present, of course, was a pale imitation of what had been wrought in the past. He made himself slave to the hope that humanity might again live up to the standards of its forebears.

His first consuming passion was for the medieval era, more specifically, for the Crusades. The Crusaders provided Lawrence a warrior ideal that would take some time for him to overcome: a knight whose heart is pledged both to an absolute ideal and an idealized love. Pining for an ideal past, Lawrence had to settle for its literary remains. With him every day of the campaign in Arabia was a masterpiece of manly striving and idealized love, Malory's *Morte d'Arthur.*

And as much as the literary, the material remains of the past held Lawrence's attention. While a student at Oxford, Lawrence began research for his thesis on the influence of eastern architecture on Crusader castles. In the summer of 1908, he completed a two-thousand-mile bicycle tour of the castles of France. It was on this tour that Lawrence first visited the cathedral at Chartres. In a letter to his mother, he attempts to find words for this seemingly mystical encounter: "[F]or it is a place absolutely impossible to imagine, or to recollect, at any rate for me: it is overwhelming . . . as though I had found a path (a hard one) as far as the gates of Heaven, and had caught a glimpse of the inside, the gate being ajar. You will understand how I felt though I cannot express myself. Certainly Chartres is the sight of a lifetime, a place truly in which to worship God."[10] On this trip came another first sight: the Mediterranean. Cresting a hill, Lawrence caught glimpse of a "silver shiver" on the horizon: "I understood, & instinctively burst out with a cry of 'θαλασσα, θαλασσα!'" "The sea, the sea!"—so Xenophon's army of the Ten Thousand shouted in relief on first sight of the Black Sea in 401 B.C., having survived a long and arduous journey back from Mesopotamia. Having experienced the present through the filter of the classical past, Lawrence made his way down to the waters of the Mediterranean: "I bathed today in the sea, the great sea, the greatest in the world: you can imagine my feelings. . . . I felt that at last the way to the South, and all the glorious East; Greece, Carthage, Egypt, Tyre, Syria, Italy, Spain, Sicily, Crete. . . . They were all there, and all within reach . . . of me. . . . Oh I must get down here—farther out—again!"[11] And indeed Lawrence did get farther out again the next summer, when he spent nearly three months on a walking tour of the Crusader castles of Syria, Lebanon, and Palestine. Here Lawrence first stepped foot in the East, in the sands of Arabia that would hold him hostage for the next ten years.

Having graduated from Jesus College, Oxford, with First Class Honors for his thesis, Lawrence was offered a position at the British Museum's excavations at Carchemish, a site he had visited on his walking tour through Syria. From the years 1911 through 1914, Lawrence and his colleague Leonard Wooley oversaw the excavation of an ancient Hittite city. Nearby was the river Euphrates, whose currents had once watered ancient Babylon. And so Lawrence, ever eager to interpret the present with the past, could have recalled Psalm 137: "By the waters of Babylon, there we sat and wept, we wept when we remembered Zion"—except that for Lawrence, his weeping would have been joyous. For it was by the waters of Babylon, in voluntary exile from England, that Lawrence passed his most blissful years. Apart from the hours digging at the site, his time was given over to hunting down Hittite seals for his mentor David Hogarth, exploring the bends of the Euphrates in his canoe, touring more of the Crusader castles of the East, and socializing with the cosmopolitan crowd that found itself in Lebanon. Among the ruins of Syria, the remains of travelers from East and West, Lawrence again caught glimpse of the divine, as he had at Chartres:

> [It was] a ruin of the Roman period which the Arabs believed was made by a prince of the border as a desert-palace for his queen. The clay of its building was said to have been kneaded for greater richness, not with water, but with the precious essential oils of flowers. My guides, sniffing the air like dogs, led me from crumbling room to room, saying, "This is jasmine, this violet, this rose." But at last Dahoum drew me: "Come and smell the very sweetest scent of all," and we went into the main lodging, to the gaping window sockets of its eastern face, and there drank with open mouths of the effortless, empty, eddyless wind of the desert, throbbing past. . . . "This," they told me, "is the best: it has no taste." My Arabs were turning their backs on perfume and luxuries to choose the thing in which mankind had had no share or part.[12]

In the hot breath of the desert, Lawrence felt a force older than history, older than the past whose remains he handled with such care. It was an intoxicating experience, this effortless emptiness, and one that he would seek out again and again.

Lover

It was a young Arab named Dahoum who led Lawrence from the perfumed rooms to drink deep of the arid nothingness of the desert. He was one of the many local villagers Lawrence and Wooley employed in their excavations. Little is in fact known about Lawrence's intimate relationship with Dahoum: a handful of photographs, a number of impersonal diary appearances, and the recollections of Lawrence's few friends who met him. Wooley recalls that "Lawrence was devoted to him." And while the Arab villagers were apparently scandalized by their unusually close friendship, Wooley, true to one manly code of his day, insists that Lawrence "was in no sense a pervert."[13] That Lawrence's love for Dahoum was never consummated is very likely: Lawrence had, by all accounts, a deep aversion to—even a loathing of—the body. As he confesses in *Seven Pillars of Wisdom* regarding "the lower creation": "To put my hand on a living thing was defilement. . . . This was an atomic repulsion. . . . The opposite would have been my choice if my head had not been tyrannous. I had a longing for the absolutism of women and animals, and lamented myself most when I saw a soldier with a girl, or a man fondling a dog, because my wish was to be as superficial, as perfected; and my jailer held me back."[14] The inhibition that prevented Lawrence from touching Dahoum may have been a combination of his puritanical upbringing at the hands of his devout but once adulterous mother, his immersion in the sublimated and aestheticized homoeroticism of Oxford, and his devotion to the warrior ideal of a medieval knight, who could never approach his idealized beloved until he completed the impossible task of earning her love.

In the conclusion to *Seven Pillars of Wisdom,* Lawrence reveals the mainspring of his action, the motivation behind the impossible task of the Arab Revolt: "The strongest motive throughout had been a personal one, not mentioned here, but present to me, I think, every hour of these two years. . . . It was dead, before we reached Damascus."[15] Dahoum had died of typhus before Lawrence could make his way north to conclude the revolt that he hoped—vainly—would win the Arabs their freedom. That Lawrence conceived the Arab Revolt as his opportunity to give Dahoum his freedom is made clear in the dedicatory poem:

> I loved you, so I drew these tides of men into my hands
> and wrote my will across the sky in stars
> To earn you Freedom, the seven-pillared worthy house,
> that your eyes might be shining for me
> When we came.

That Lawrence announces that his writing is both an offering to, and takes its origin from, a beloved places him in a long tradition of eminent writers for whom erotic yearning for another is inseparable from their art. That Lawrence thereby takes his place beside Dante and Shakespeare is made all the more poignant when we remember that Dahoum, like Beatrice, is no longer among the living, and like the young man of the Sonnets is less cultivated than his admirer.

Some will object that I have conflated Dahoum and "S. A.," to whom the poem—and thereby the book and the revolt—is properly dedicated. Many scholars have thought to dispel the mystery surrounding the identity of S. A., but Lawrence, swinging between disclosure and concealment, has confused their efforts with indirection, irony, and half-truths. Very few deny that Dahoum is at least some part of the identity of S. A. Indeed in his correspondence to his biographers, Lawrence suggests that the "A" of S. A. stands for a personal motivation, but that the "S" stands for either an idea or a place.[16]

Sometimes, as in this note to his biographer B. H. Liddell Hart, Lawrence highlights the one sense over the other: "S. A., the subject of the dedication, is rather an idea than a person."[17] That Lawrence would never definitively articulate the senses of S. A. should not surprise us; he was, after all, a "boyish leg-puller" who smiled when his allusions exhausted the efforts of the scholiasts. Nor should it surprise us that his personal motivation, his love for Dahoum, was rarefied into an ideal, even a place. It is of a piece with his romantic idealism, freedom for the East as a gift for his beloved. Here in one formula is ideal, place, and person.

Man of War

Colonel T. E. Lawrence was never a soldier. He made his way to the field by way of the desks of the Intelligence Office in Cairo, and so escaped formal military training. He revered the disciplined but elastic mind of General Allenby, who would lead the British forces north through Gaza, Beersheva, and Jerusalem to Damascus. He basked in the glory of the Homeric Auda Abu Tayi, the Howeitat chief who married twenty-eight times, had been wounded thirteen times, and had himself slain seventy-five men—all of them Arabs—with his own hand in battle. Of Turks killed Auda kept no record, esteeming them so low. Lawrence's great gift to the Arab Revolt was not the disciplined mind of a soldier or the berserk rage of a Howeitat chief, but the mind of a gifted strategist.[18]

In its first movements, the Arab Revolt had succeeded in freeing Mecca from the Turks. But the Turkish garrison in Medina, the city second holiest to Islam, had repelled the Arab advance. In their next, very precarious steps, the Arab forces eventually surrounded the garrison at Medina so that while the Turks could not be driven out, neither could they spare troops to pursue the rebels. It came as a great surprise to the Arabs when the British asked that this garrison at Medina not be allowed to retreat north, that Medina not be liberated. It was left to the liaison officer Lawrence to explain to Prince Feisal, one of the sons of the emir of Mecca and perhaps the principal leader of the Arab Revolt, that the British request was in the Arabs' own best interests. Were the Turks at Medina to retreat north, they would join forces with the garrison at Maan, preparing to meet Allenby's advance from the southwest. When the Arabs themselves wished to move the revolt north, they would meet these same Turks again, refortified. Nor could the Arabs rely on the British, for who could be sure that they would prevail against such a Turkish force, doubled in size?

These and other epiphanies seized Lawrence in a delirious state of illness, a week-long bout with dysentery. He understood that the British request must be answered, and yet that the Arab Revolt must also move north if it was to win freedom for greater Syria as well. The fire was stoked with the knowledge, which he now shared with Feisal, that the British and the French had formal designs on the Near East, the French in the North and the British in the South, and that the earlier promises of Arab independence would be smoke if they did not push the revolt north and take Damascus.

But first the Arabs needed to take Aqaba. Aqaba was a port town from which the Arab Revolt could be easily supplied by the British navy. But the mountain passes that led northward from Aqaba were too heavily fortified to charge head on from the shore. And so Lawrence, delirious with dysentery, conceived of a decorous approach, a backdoor, for "in the East they swore that by three sides was the decent way across a square."[19] They would approach the passes to Aqaba from inland and surprise the Turks from behind. All that lay between them and their surprise was the three sides of the square: hundreds of miles of the harshest deserts in Arabia. Lawrence and Auda, together with Sherif Nasir, set out with a small party on their errand to test themselves and God. And after their forty years in the Sinai, they too were delivered into Zion. Nasir and Auda were able to enlist the help of other tribes, and with this substantial force of irregulars they stormed the mountain passes and captured Aqaba. It was the first great and, to the British, totally unexpected victory of the Arab Revolt.

With the capture of Aqaba, the scope of the Arab Revolt changed dramatically. No longer could it be an independent and haphazard band of brigands, slaves to the ideal of freedom. For the Arab Revolt to survive and grow, it must become the right flank of Allenby's regular army. This required sensitive diplomatic maneuvers, including having Feisal acknowledge Allenby, however nominally, as the commander of all operations. Between Cairo and Aqaba, the British and the Arabs, Lawrence became the film through which information passed, the mind by which crucial translations—linguistic, diplomatic, and conceptual—were made. Here again are his two selves, the poles between which his being swung: "Sometimes these selves would converse in the void; and then madness was very near, as I believe it would be near the man who could see things through the veils at once of two customs, two educations, two environments."[20]

The Arab Revolt was to become what Lawrence would later call "a side show of a side show,"[21] a crippling nuisance to the Turks in the East: "an influence, an idea, a thing intangible, invulnerable, without front or back, drifting about like a gas . . . so we might offer nothing material to the killing."[22] The Arabs would suddenly appear from and disappear into the desert, unpredictably. With these unsettling maneuvers they succeeded in keeping the Turks occupied and agitated. And when it came time for Allenby's great advance, the Arabs could so pepper the Turks' precious railway with mines that they could not retreat or regroup. When that time came, in

September 1918, Allenby and the British and Lawrence and the Arabs crushed the Turks as between a hammer and an anvil, and both armies rushed north to claim the prize of Damascus.

Diplomat

Few would envy Lawrence's diplomatic situation.[23] He took upon himself the responsibility for negotiating between a schizophrenic British foreign policy and the nascent Arab aspirations for independence, represented most visibly by the dynastic interests of the Hashemites in Mecca.

In order to rouse the sleeping giant of Arabia, the British had to make certain promises. Shortly after the outbreak of the war, Sir Henry McMahon, the British High Commissioner in Cairo, opened a correspondence with Sherif Hussein, the emir of Mecca, in which he promised that were the Arabs to revolt against the Turks, they would be guaranteed their independence. This appealed to Hussein, who had designs to make himself king of all Arabia, from the Hijaz to Syria. But the British had commitments to another, more powerful ally: the French, who claimed that their stake on Syria and Lebanon went as far back as the Crusades. With the Arabs, the British negotiated along the lines of the Hussein-McMahon Correspondence; with the French they operated according to the Sykes-Picot Agreement, in which the French were granted custodial rights over Syria and Lebanon, the British over Palestine, Jordan, and much of Iraq.

Lawrence learned of the Sykes-Picot Agreement in May 1916, while he was still working in Cairo. The ineptitude—a generous interpretation—or dishonesty—a less generous one—of the British policies appalled him. But with a true diplomat's eye, he saw in the wording of the agreement a window of possibility: were the Arabs to capture Homs, Hama, Aleppo, and, most importantly, Damascus, they would have a legitimate claim to autonomy in Syria. As Lawrence wrote later, in the Introductory Chapter of *Seven Pillars of Wisdom:*

> I salved myself with the hope that, by leading these Arabs madly in the final victory I would establish them, with arms in their hands, in a position so assured (if not dominant) that expediency would counsel the Great Powers a fair settlement of their claims. In other words, I presumed (seeing no other leader with the will and power) that I would survive the campaigns, and be able to defeat not only the Turks on the battlefield, but my own country and its allies in the council-chamber. It was an immodest presumption.[24]

Lawrence shared this bold hope with Feisal in order to convince him to leave Medina behind, to push north toward Aqaba and Damascus. When, on October 1, 1918, Arab forces did enter Damascus, Lawrence's hope was still alive.

Almost immediately, Lawrence departed for England, leaving Feisal in somewhat nominal control of Damascus. For the remainder of the year, Lawrence lobbied for the Arabs in London, but to little avail. It became clearer to him just how much of a side show of a side show his campaign and now his cause were. The British were too interested in appeasing France and securing their own colonial territories in the Near East to care much for fulfilling McMahon's promises to the Arabs. Lawrence did succeed in securing for Feisal a place at the Paris Peace Conference in 1919 and for himself a place in the British delegation attached to Feisal's staff. For Feisal he was translator and advisor, negotiating the diplomatic circles of Europe as he had the tribal meetings of Arabia. But he found that his skills of persuasion were less effective on his own. Nor was he above playing the exotic in order to win attention to his cause. He and Feisal would arrive at formal diplomatic events in full robes and headdresses, and so Lawrence contributed to the romantic myth of "Lawrence of Arabia" that he would later struggle to escape. Despite the pair's celebrity, Feisal's ambitions in Syria and the wider question of Arab independence were met with little more than vague sympathies. Little was decided at Paris regarding the fate of Arabia, and both Lawrence and Feisal left Paris uncertain and dejected.

In the next two years, the face of the Near East changed dramatically. Britain was awarded mandates over Iraq, Transjordan and Palestine, France over Syria and Lebanon. Sherif Hussein of Mecca was driven out of the Hijaz by Ibn Saud, and the Arabian Peninsula was consolidated under his rule as Saudi Arabia. Feisal was declared king of Syria by a national congress; this prompted the French to invade Syria and establish their own unambiguous rule there. In Mesopotamia in southern Iraq there was an armed rebellion against British governance, which the British severely suppressed.

Into this quagmire stepped Winston Churchill, Colonial Secretary, who invited Lawrence to join his department as an advisor. In March 1921, Churchill called a conference in Cairo to settle the burning questions of the Near East. Palestine was to remain a British protectorate, to become the battleground between British, Arab, and Zionist ambitions. The rule of Transjordan was given to Prince Abdullah, son of Sherif Hussein and brother of Feisal. There to this day the Hashemite kingdom reigns. The crown of Iraq was offered to Feisal as a sort of consolation prize for losing Syria. The sham that many of these decisions may seem to us now were to Churchill and Lawrence a satisfactory solution to a volatile situation. To Robert Graves, Lawrence later wrote: "I take most of the credit for Mr. Churchill's pacification of the Middle East upon myself. I had the knowledge and the plan. He had the imagination and the courage to adopt it and the knowledge of the political procedure to put it into operation. . . . [The settlement was] the big achievement of my life: of which the war was a preparation."[25] In *Seven Pillars of Wisdom,* he seems to absolve himself and the British from any further culpability: "[A]t his conference in Cairo . . . [Churchill found] solutions fulfilling (I think) our promises in letter and spirit (where humanly possible) without sacrificing any interest of our Empire or any interest of the peoples concerned. So we are quit of the wartime Eastern adventure, with clean hands, but three years too late to earn the gratitude which peoples, if not states, can pay."[26]

These passages, especially the second, read as if Lawrence were trying to convince himself of the good results of his diplomatic efforts. For he knew as well as anyone that nation-states, Hashemite monarchs, and British tutelage were no more native to the Near East than Turkish governance. That Lawrence's regard for the interests of Empire was in most instances negligible, even hostile, suggests that this note to the main text had a certain audience in mind, both an audience within himself wishing to be freed of the burden of personal shame, and an audience among British readers hoping to have explained to them that Empire had indeed secured peace with integrity. Many today would prefer the honest note Lawrence strikes more often in *Seven Pillars of Wisdom:* "Clearly I had no shadow of leave to engage the Arabs, unknowing, in a gamble of life and death. Inevitably and justly we should reap bitterness, a sorry fruit of heroic endeavour."[27]

Writer

In *Seven Pillars of Wisdom,* Lawrence confesses, "I had had one craving all my life—for the power of self-expression in some imaginative form—but had been too diffuse ever to acquire a technique."[28] His ambition for self-expression found scope even before the war, in Carchemish. There he conceived plans to write a history of the Crusades, a life of Christ, and an account of his travels through seven cities of the Near East, to be entitled *Seven Pillars of Wisdom.* On this last endeavor he made real progress, but he destroyed the manuscript before the war in 1914. While in Paris in 1919, he found that he had a considerable amount of idle time between his frustrated efforts to lobby for Feisal, and so he gave in to his old craving for self-expression. Over the next seven years, he would craft an account of the Arab Revolt to which he attached the same title as his aborted travel narrative.

Seven Pillars of Wisdom is a book that, much like its author, confounds categories. It is, most simply put, a narrative of the course of the Arab Revolt and Lawrence's substantial part in it. But around "this tent-pole of a military chronicle," E. M. Forster writes, "T. E. has hung an unexampled fabric of portraits, descriptions, philosophies, emotions, adventures, dreams."[29] The diffusion that, Lawrence suggests, inhibited his acquiring a technique for self-expression, instead became a manifold technique for self-expression. The mammoth literary result is, no less than its Odyssean author, polytropic. Many critics have sought to answer the question of the genre of *Seven Pillars of Wisdom:* is it a history, an autobiography, a war memoir, a confession, an epic, a tragedy, or a triumph?[30] In a letter to Edward Garnett, Lawrence expressed his hope that *Seven Pillars of Wisdom* might be a "Titanic" book, "those distinguished by greatness of spirit."[31] "Titanic" was a category in which Lawrence included Dostoevsky's *The Brothers Karamazov,* Nietzsche's *Thus Spoke Zarathustra,* and Melville's *Moby Dick.* That these books are "Titanic" suggests that they are immense, motivated by a power impossible to control, of a mythic era that is long past, and, because they attempt more than is their due, doomed to failure, destined to fall short of their impossible goals.

Although his literary friends praised it to no end, and generations of readers have fallen under the spell of its hauntingly archaic prose, *Seven Pillars of Wisdom* was, like the events it narrates, a great disappointment to Lawrence. His other great venture into the dangerous game of creation is *The Mint,* a shockingly honest account of his life in the ranks of the RAF. In style and structure nothing like *Seven Pillars of Wisdom, The Mint* has the sparse economy and candid immediacy of a diary. But even in this new register, Lawrence's powerful descriptive skills, penetrating character sketches, and passionate arias are still recognizable.

I opened this introduction with a discussion of Lawrence's translation of Homer's *Odyssey,* one commissioned by Bruce Rogers. But how does the art of translation relate to Lawrence's life as a writer? In one of his many letters to Rogers, Lawrence explains his unique qualifications for this project: "For years we were digging up a city of roughly the Odysseus period. I have handled the weapons, armour, utensils of those times, explored their homes, planned their cities. I have hunted wild boars and watched wild lions, sailed the Aegean (and sailed ships), bent bows, lived with pastoral peoples, woven textiles, built boats and killed many men. So I have odd knowledges that qualify me to understand the *Odyssey,* and odd experiences that interpret it to me."[32] We might add that the *Odyssey* too has odd knowledges and odd experiences, ones that helped Lawrence interpret himself and now help us interpret him. For although Lawrence came to regard translation as a mechanical art, an escape from the vulnerable venture of literary and self creation, I have included it here under the rubric of writer rather than engineer (where he might have put it) because of a conviction, shared by at least two of the contributors to this volume, that Lawrence was wrong, that his translation of Homer's *Odyssey* is in fact a remarkable instance of literary and self-creation. When the past is a filter through which to experience the present, as it was for Lawrence, then archaeology can be a mode of self-discovery and translation a mode of self-expression.

Engineer

In 1922, Lawrence did what few would expect of a national celebrity and military hero: he enlisted in the ranks of the RAF under an assumed name and disappeared, for a time, from the public eye. And despite a brief discharge due to press exposure, during which Lawrence enlisted in the Tank Corps, he would remain in the ranks of the RAF until February 1935, just a few months before his death. What might have motivated this peculiar move? In a word: release. Release from the responsibility he bore for the Arab Revolt's success and failure; release from the shame of what had come of those efforts in the wake of Paris and Cairo; release from his service to two worldly masters, British and Arab; release from the shackles of his own legend, Lawrence of Arabia, which he both promoted and loathed; release from his knightly quest to earn the love of his beloved—for all of these reasons Lawrence sought to slip away into anonymity. In the ranks of the RAF, he thought he might learn to be an ordinary man and not to dream with open eyes, as he had in Arabia, but to do as he was told. He sought a more modest scope in which to rebuild his shattered self. If in Arabia he wrote his will across the sky in stars, in the RAF he would train his will and body to perform drills on the parade ground. Lawrence of Arabia was to become 352087 A/c Ross.[33]

In the RAF, Lawrence became a mechanic, both in letter and spirit. In Arabia, Lawrence had taken a boyish glee in the mechanisms of the demolitions he and the other British officers had developed to wreck such havoc on the Turks' railway. But now, in the engines of airplanes and speedboats, Lawrence found a different pleasure. Machines, while dynamic, were orderly. There was no morality in the machine, only power and efficiency, both of which could be improved upon, with clear conscience, by the creative mind. Such improvements were well received by his superiors, and Lawrence never feared that the good work he set his mind and hands to do would be wrecked by the narrow thinking of old men.

The elements he devoted himself to in those years were air and water. Besides the planes, Lawrence spent time developing high-speed rescue boats for the RAF. Suspended in the air, above the world, or rushing over the sea in a boat, Lawrence would come to experience a certain bliss, the slaking off of the tormented self, the oblivion, for a moment, of his many ways: "We in the machine were like souls suspended motionless in unchanging ether, conscious of no movement, of no space, hardly of time—for comparing notes with one another afterwards we could not rightly say if the four hours of our crossing had seemed to us a moment or an age. For that space our minds had ceased to exist."[34] In the air as upon the water, Lawrence would come to feel the breath of that same intoxicating wind, that effortless emptiness, that he had first tasted with Dahoum in a ruined castle of Arabia. It was in the desert, as in the air and on the water, that Lawrence would come to worship what he called in *Seven Pillars of Wisdom* a "god of negations."[35]

But the experience to which Lawrence was most addicted in his last years was speed. And the machine that fed this addiction was his motorcycle, Boanerges, "sons of thunder." It was appropriate that Lawrence bestowed upon this steed the name that Jesus gave his disciples James and John, for he came to regard this machine as a loyal friend: "An almost human machine, he is, a real prolongation of my own faculties: and so handsome and efficient. Never have I had anything like him."[36] Racing with Boanerges on country roads, Lawrence could have a taste, on the ground, of some of the effortless emptiness he experienced in the air or on the water. On Boanerges, Lawrence could outrun himself. When he caught up with himself, however, he often thought that his efforts were little more than a sad game: "When my mood gets too hot and I find myself wandering beyond control I pull out my motor-bike and hurl it top-speed through these unfit roads for hour after hour. My nerves are jaded and gone near dead, so that nothing less than hours of voluntary danger will prick them into life: and the 'life' they reach then is a melancholy joy at risking something worth exactly 2/9 a day."[37]

On May 13, 1935, his risk became a reality. When he died six days later from a head injury, few would agree with him as to the worth of his life. Perhaps he was following his own Epicurean advice, with which, with the same "melancholy joy" as he found on Boanerges, he once consoled his mother over the death of his younger brother, Frank: "You know men do nearly all die laughing, because they know death is very terrible, & a thing to be forgotten till after it has come."[38]

When will Lawrence be forgotten? Will he soon fade from our imagination? Some have thought that he would have disappeared by now, but his legacy has endured through a new century, despite his detractors' and even, at times, his own best efforts. His influence is difficult to register today: it can no longer be measured in grand movements or institutions, but only in individuals, here and there, who have fallen under what B. H. Liddell Hart calls the "spell" of his personality.[39] He has become, like the Arab Revolt, "an influence, an idea, a thing intangible, invulnerable, without front or back, drifting about like a gas . . . a vapor."[40] It is not surprising that this gas has, over time, collected in certain pockets: more writers than academics have been interested in Lawrence, more explorers than historians. Our ambivalence to the word "spell" helps explain both why some dismiss him as a charlatan and others revere him as a quintessentially modern prophet, one who dons "doubt, our modern crown of thorns."[41] A similarly ambivalent reception has been given Odysseus from antiquity forward. And Odysseus has also been thought, by every age, to be the quintessentially modern hero. Perhaps then

Lawrence's fate is to be much like Odysseus': both polytropes are uneasy heroes and poets whose peculiar personalities speak afresh to every generation. Perhaps the pair are to remain, in our imagination, ever ancient and modern.

Notes

1. Letter to Bruce Rogers dated February 1, 1928 (from Introduction by Bruce Rogers to *Letters from T. E. Shaw to Bruce Rogers,* unpaged), quoted in John E. Mack, *A Prince of Our Disorder: The Life of T. E. Lawrence* (Boston: Little & Brown, 1976), 373. When Lawrence enlisted in the Tank Corps in 1923, he did so under the name "Shaw." He kept that name when he reenlisted in the RAF in 1925.
2. T. E. Lawrence, trans., *The Odyssey of Homer* (New York: Oxford, 1991), xxiii. For a more detailed discussion of Lawrence as an Odyssean *polytrope,* see Maren Cohn's essay in this volume.
3. Stephen E. Tabachnick, ed., *The T. E. Lawrence Puzzle* (Athens: University of Georgia, 1984), 1.
4. Robert Graves and B. H. Liddell Hart, eds., *T. E. Lawrence to His Biographers* (Garden City: Doubleday, 1963), Vol. I, 63.
5. André Malraux, "Lawrence and the Demon of the Absolute," *Hudson Review* 8 (1956).
6. Letter to Eric Kennington, dated August 6, 1934, David Garnett, ed., *The Letters of T. E. Lawrence* (Garden City: Doubleday, 1938), 813–14.
7. A. W. Lawrence, ed., *T. E. Lawrence by His Friends* (London: Cape, 1954), 82.
8. Ibid., 165.
9. Letter from Fareedah el Akle to Helen Cash, dated June 12, 1969, quoted in John E. Mack, *A Prince of Our Disorder,* 78.
10. Letter to Sarah Lawrence dated August 28, 1908, Malcolm Brown, ed., *T. E. Lawrence: The Selected Letters* (New York: Paragon, 1992), 16–17.
11. Ibid., 15.
12. T. E. Lawrence, *Seven Pillars of Wisdom* (London: Penguin, 1962), 38.
13. A. W. Lawrence, *T. E. Lawrence by His Friends,* 75.
14. T. E. Lawrence, *Seven Pillars of Wisdom,* 580–81.
15. Ibid., 684.
16. Robert Graves and B. H. Liddell Hart, eds., *T. E. Lawrence to His Biographers,* Vol. II, 143.
17. Ibid., 55.
18. Readers interested in Lawrence's strategic thinking should consult George W. Gawrych's essay in this volume.
19. T. E. Lawrence, *Seven Pillars of Wisdom,* 63.
20. Ibid., 30.
21. T. E. Lawrence, "Evolution of a Revolt" in S. and R. Weintraub, eds., *Evolution of a Revolt: Early Post-War Writings of T. E. Lawrence* (University Park

and London: Pennsylvania State University Press, 1968), 118; quoted in Jeremy Wilson, *Lawrence of Arabia* (Stroud: Sutton, 1998), 55.

22. T. E. Lawrence, *Seven Pillars of Wisdom,* 198.
23. This discussion owes much to Jeremy Wilson's very helpful shorter biography of Lawrence, *Lawrence of Arabia.* See especially chapters two and five. Specifically, I owe him the metaphor of the hammer and anvil. His longer biography provides an exhaustive account of Lawrence's life, *Lawrence of Arabia: The Authorized Biography* (New York: Atheneum, 1990).
24. T. E. Lawrence, *Seven Pillars of Wisdom,* 24.
25. Robert Graves and B. H. Liddell Hart, eds., *T. E. Lawrence to His Biographers,* Vol. I, 112. Quoted in Jeremy Wilson, *Lawrence of Arabia,* 81–82.
26. T. E. Lawrence, *Seven Pillars of Wisdom,* 283n.
27. Ibid., 283.
28. Ibid., 565.
29. E. M. Forster, back cover of the Penguin edition of *Seven Pillars of Wisdom* (London: Penguin, 1962).
30. In this volume alone, Theoharis presents it as a triumph, Cohn as an anti-epic, and I present it as a history on the model of Herodotus.
31. David Garnett, *The Letters of T. E. Lawrence,* 360.
32. Letter to Bruce Rogers, dated January 31, 1931, in David Garnett, ed., *The Letters of T. E. Lawrence,* 710.
33. Lawrence first enlisted in the RAF under the name "J. H. Ross." When he enlisted in the Tank Corps in 1923, he used the name "T. E. Shaw." When he was readmitted to the RAF in 1925, he kept the name "Shaw" and became 338171 A/c Shaw.
34. From T. E. Lawrence, "A Sea Trip Essay," dated April 18, 1923; quoted in John E. Mack, *A Prince of Our Disorder,* 342.
35. T. E. Lawrence, *Seven Pillars of Wisdom,* 39.
36. Letter to C. F. Shaw, dated August 24, 1926, quoted in Jeremy Wilson, *Lawrence of Arabia: The Authorized Biography,* 772.
37. Letter to L. G. Curtis, dated May 14, 1923, quoted in ibid., 719.
38. Letter to his mother, undated [1915], in Malcolm Brown, ed., *T. E. Lawrence: The Selected Letters,* 73.
39. B. H. Liddell Hart, *T. E. Lawrence in Arabia and After* (London: Jonathan Cape, 1935), 18; quoted in Paul Marriot and Yvonner Argent, eds., *The Last Days of T. E. Lawrence* (Portland: Alpha, 1996), 168.
40. T. E. Lawrence, *Seven Pillars of Wisdom,* 198.
41. Ibid., 36.

Chapter One

"I had had one craving . . ."

Theoharis Constantine Theoharis

Mild consternation must often have afflicted browsers looking for T. E. Lawrence's *Seven Pillars of Wisdom* in bookstores. Without the kind of help provided by overqualified clerks in such academic enclaves as Cambridge or Hyde Park, the aspirant reader may very well travel from the autobiography section of the store to the history section and then to the gay and lesbian section and then to the Middle East section only accidentally to see the book shelved with other memoirs. The classification problem that Lawrence's masterpiece poses for those engaged in literature as a commercial venture and for those pursuing reading as an edifying mode of leisure is, admittedly, trivial. But for those who read to enhance or discover a mode of life—for those, in other words, who read in the great tradition Lawrence clearly appeals to throughout this book—its unusual formal structure, the problems of classification it poses, are not trivial vexations but extraordinary calls to transcendence. On its title page Lawrence christened the book "a triumph." What would it mean to take him at his word, to think of *Seven Pillars of Wisdom* as not only about triumph, but to think of it as a new genre called "a triumph," a new literary form that imaginatively instantiates, and so permanently makes known, whatever value Lawrence thought "triumph" gave to life?

Twentieth-century writers have borrowed forms from other arts to signal the innovation aspired to in their work; T. S. Eliot's *Four Quartets* comes immediately to mind. But the transposition of non-aesthetic structures into literary structures is rare. Indeed, Lawrence's turning to classical history for the formal, generic name of *Seven Pillars of Wisdom* is probably a unique instance of a modern author grafting an ancient martial and political ceremony onto the personal account of a modern war or any other subject, fictional or not. The subtitle, for all its strangeness (and perhaps because of it), may easily slip past readers, vanish as a merely "literary" opening flourish once they push forward through the story of the campaign. But after thinking of or learning what a "triumph" is, those same readers may also stop at the term and sense themselves invited and instructed by it to experience a special, new significance in the elevated pitch and extensive range of feeling in which Lawrence tells the story of his part in the Arab Revolt in World War I.

Ernst Badian, in *The Oxford Classical Dictionary,* enticingly subtitled *The Ultimate Reference Work on the Classical World,* offers the following account of the classical allusion on Lawrence's title page.

> triumph, the procession of a Roman general who had won a major victory to the temple of Jupiter on the Capitol. . . . In classical times, the procession entered Rome through the *porta triumphalis* ("triumphal gate") through which no one else might enter. . . . It made its way to the Capitol by a long route including open spaces where large numbers could see it. It comprised, essentially, the *triumphator* (dressed in the costume said to have been the kings' and close to Jupiter's) on a four-horse chariot, with any sons of suitable age as outriders; eminent captives (normally destined for execution) and freed Roman prisoners of war dressed as the *triumphator's* freedmen; the major spoils captured; his army; and animals for sacrifice. The whole senate and all the magistrates were supposed to escort it. Increasingly costly and elaborate details were added from *c.* 200 B.C., including banners, paintings of sieges and battles, musicians, and torch-bearers. The *triumphator* was preceded by his lictors, and a slave rode with him, holding a laurel wreath over his head and reminding him that he was mortal. The soldiers chanted insulting verses, no doubt to avert the gods' displeasure. . . . The right to triumph depended on a special vote of the people allowing him to retain his military *imperium* in the city, and so in fact on the senate's decision to ask for this vote.
>
> In classical times the prerequisites for expecting such a decision were a victory in a declared war over a foreign enemy, with at least 5,000 of them killed and the termination of the war; . . . In the late republic interpretation of the rules came to be dominated by power and influence. . . . [1]

A plenary civic celebration and ornamentally symbolic narrative of victory in war, a religious rite and a propaganda parade that enact the hero's championing of and service to absolute divine and state power, a telling by showing of the conqueror's adequacy and inadequacy to the magnificently accomplished task—these, among other elements of the classical Roman triumph, offered Lawrence a splendidly rich and complex form for his tale (he calls it a tale in the book's first sentence) of disappointing success. That he so quickly switches from "triumph" to "tale" ironically gives a grand thematic scope to the circumstantial evil of the war that he refers to in the opening paragraphs. Fighting in the extreme conditions of the desert, Lawrence reasons, exacerbated a disastrous moral difficulty already inherent in serving Ideal Freedom.

> As time went by our need to fight for the ideal increased to an unquestioning possession, riding with spur and rein over our doubts. Willy-nilly it became a faith. We had sold ourselves into its slavery, manacled ourselves together in its chain-gang, bowed ourselves to serve its holiness with all our good and ill content. The mentality of ordinary human slaves is terrible—they have lost the world—and we had surrendered, not body alone, but soul to the overmastering greed of victory. By our own act we were drained of morality, of volition, of responsibility, like dead leaves in the wind.[2]

Passages like this, where moral failure and anguish reward the aspirant fighter, abound in *Seven Pillars of Wisdom.* In light of them, where conquerors are made slaves, what prevents readers from thinking that Lawrence's formal, titular presentation of the book as a "triumph" should be thought of as anything more than learnedly sarcastic self-abasement? What, if anything, of the conquering Roman victor's ceremoniously ambiguous glory does Lawrence's tale preserve for him and for his readers?

Costly and elaborate ornamentation of the military narrative, for one thing, including paintings. The Roman triumphal sequence was a decorated pageant symbolically narrating the war's history—hence the inclusion in it not only of the conquering general and his army, but also of the captives and the ceremoniously attired representatives of the victorious state for which the general and army fought. Lawrence took great pains to ensure that *Seven Pillars of Wisdom* would be physically produced with every available aesthetically accomplished elaboration of the bookmaker's craft, from rich leather binding, heavy paper, and original lettering, to dozens of specially commissioned paintings and drawings of the campaign's Arab and European principals. The insistence on his book's rare and sumptuous physical beauty had a moral and cultural meaning for Lawrence, who, through such devotion to the material craft of literature, took his place in the Aesthetic Movement launched in Victorian England by William Morris, one of the idols of Lawrence's youth. Through his magnificently produced Kelmscott Press books, through his production of furniture and decorative furnishings at Morris & Co., and in the idealized heroism of his medieval romances and his original verse and fiction, Morris explicitly protested what he and many other Victorians, John Ruskin also comes to mind, regarded as the vulgar uniformity of sentiment, thought, and value in the rapidly ascendant industrial age, whose chief offense was mechanization and hence commodification of human striving in the material world. Quixotic as this assertion of the creative individual was, it became part of the anti-modernism of the great literary modernists in the British Isles who were Lawrence's contemporaries, especially Yeats, and ironically, Joyce. And it engaged those Americans who were in exile in England during and after the World War I as well, including Eliot and Pound, who made his own furniture for a while.

The Aesthetic Movement's idealized championing of the individual's effort to make life freely beautiful, carried out against the totalizing worldly forces that had made life slavishly ugly, amounted to a highly subjective cultural politics as well as a quixotic war against materialism. Lawrence bountifully and exquisitely channeled those forces into his modernist Roman triumph. The war Lawrence writes of in his triumph is not only a subjectively cultural one, though, but a soldierly, real world event, for all the adventure storying and medieval romancing Lawrence lards into his telling of the Arab campaign. To put this observation more abstractly, the idea and the fact of materialism figure prominently throughout *Seven Pillars of Wisdom,* primarily as the twinned forces that Lawrence's idealization of freedom never successfully mastered or obeyed. Morris's aestheticism gives a valorous sheen to this exquisite imbalance in Lawrence's tale, and points, as beauty at its best does, to a truth. The truth in this case, discordantly displayed in the Roman triumph, is that power is never absolute or absolutely wielded. A nuance in the ancient celebration of military success, this truth dominates everywhere in the material, conceptual, political and moral soldiering Lawrence recounts in his own discordant "triumph."

Dressed as Jupiter himself, the conquering general, escorted by the whole government of Rome on his triumphant way to the Capitol, was jeered at by his army. While he listened to cheers and trumpets on his progress to worship the almost omnipotent god whose costume he assumed, the victor continually heard a slave, who held a symbolic crown of laurel over him, whisper in his ear that he would die. As religion, the jeers and whispers amount to little more than superstitious propitiation of Roman divinity. But that whispering slave's publicly comic and privately (if anything is private in a parade) solemn lip service to the supernal powers incidentally certifies the hero's piety and so magnifies his manly virtue. And finally, as politics, the mocking that humbles the hero deftly amplifies his exemplary status as loyal public champion. These crafty Roman devices of glorification through minor inversions of the conqueror's power are mightily extended in the modern Englishman's depiction of his Arab triumph. There the jeering inversion of champion and servant and the public performance of a normally private *memento mori* meditation become weighty moral inquiries into Lawrence's spiritual and political dilemmas, and, at their gravest and most moving, tragic presentation of the private and extraordinarily erotic aspiration underlying both.

The inversions of glory in the Roman triumph were meant to check whatever craving for exalted self-expression that ceremony would naturally induce or enhance in the conquering hero. This is in keeping with the pre-Imperial Roman ethos that required magnanimity of its best men but severely censored any display of personal ambition in them, judging such self-expression dangerous and corrupting betrayal by one man of the higher good of the Senate and the People of Rome whom he served. (One thinks of Julius Caesar.) In complex ways Lawrence used the triumph's chastening of the hero it exalts to express precisely that personal ambition in heroism that the ceremonious Roman form sought to suppress. His version of the triumph's chastening of the private man centers on two private imperatives for action to which *Seven Pillars of Wisdom* is confessional testimony. First, the drive for self-creation through arduous service to the beloved, and second, the craving for self-expression.

> I loved you, so I drew these tides of men into my hands
> and wrote my will across the sky in stars
> To earn you Freedom, the seven pillared worthy house,
> that your eyes might be shining for me
> When we came.[3]

These justly famous lines announce, in the poem that starts the book, that Lawrence's campaign in the desert was a love tribute, a knight's erotic exploit dared for the endangered ruler of his heart. The mastering beloved in this case was Dahoum, a young Arab whom Lawrence first knew as a servant at a dig in Carchemish in Syria in 1911. The poem, which disguises the beloved's identity in the dedicatory initials S. A., goes on to say that after Dahoum's death Lawrence shattered the house Freedom would have made for the beloved—i.e., that Lawrence left the campaign unfinished. The abandoned political project offers the Arabs who remain not the seven-pillared house that would have been Dahoum's, but its ruin, in whose "marred shadow" they may "patch themselves hollows."[4]

In fact Lawrence accomplished his military objective in the revolt—to successfully get the Arabs to Damascus—and strove to secure independence for them at the Paris Peace Conference that followed World War I. But he had also served English and French imperialist ends in both endeavors, and felt himself compromised by that double agency throughout the war and the negotiations. Lawrence's poetic shattering of the house of Freedom metaphorically presents his disappointed abandonment of any hope that the war would end, for the Arabs, in that freedom he and they had undertaken it to achieve. The same metaphoric shattering also bitterly depicts what Lawrence thought of as his own betrayal of the Arabs' hope through his service to European powers that had cynically exploited that hope from the first. In these senses, while the poem magnifies that hero's private action the triumph would expunge, it also shows that personal ambition chastened, as the triumphal form requires. Such are the antique ironies by which Lawrence ceremoniously acclaims that idealized active life he errantly undertook in the desert, that waking dream by which he sought to love and free an Arab lad.

The monumental house Lawrence shattered instead of finishing is also the book to which the poem is prologue. In this case the shattering metaphorically depicts authorial disappointment over what Lawrence regarded as the book's diffuseness and his own untimeliness as its writer. He puts the matter as follows:

> It was a hard task for me to straddle feeling and action. I had had one craving all my life—for the power of self-expression in some imaginative form—but had been too diffuse ever to acquire a technique. At last accident, with perverted humour, in casting me as a man of action had given me place in the Arab Revolt, a theme ready and epic to a direct eye and hand, thus offering me an outlet in literature, the technique-less art. Whereupon I became excited only over mechanism. The epic mode was alien to me, as to my generation. Memory gave me no clue as to the heroic, so that I could not feel such men as Auda in myself.[5]

In the excerpted authorial self-portrait, the chastening from the Roman triumph appears in Lawrence's censoring of the motive, the occasion, and the result of his self-expression in this narrative of the Arab Revolt as all inadequate.

With "craving" Lawrence casts a pejorative pall over his one lifelong drive, naming it an inordinate physical desire and thus implying that an unmanly incontinence had always governed his life of feeling and finally corrupted his ambition to find scope for that life in imaginative self-expression. The taint of disordered physicality indelibly marking each level of sublimating idealization in heroic action is a theme Lawrence sounds throughout the book; it comes as no surprise, then, that he so manfully reveals the autobiographical source of that taint. And manliness is the issue, specifically the manly talent for disposing the force of feeling in the life of action, that Lawrence critically probes in himself here. The critique involves a minimally disguised allusion to Hamlet's version of that problem. More than an idealizing literary flourish, the reference brings to light, for those alert to it, that innocent incrimination, that confession and complaint of inadequacy, which comes with any disclosure of the disappointed self.

Lawrence presents the personal origin for his writing on a page he titled "Out of Place." He could have added "out of time," as the following indicates: "At last accident, with perverse humour, in casting me as a man of action had given me place in the Arab Revolt. . . ."[6] That "casting," both dramatic assignment and dismissive play on Fortune's part, provided Lawrence the epic material for the epic writing that, had he been a different man in a different time and place, would have satisfied his lifelong craving for self-expression. Hamlet similarly scrutinizes his alienation once he is called to action, observing, as Lawrence does here, that the call is a cruel cosmic joke by the forces of time and place at his expense. His last words once he has taken up the ghost's commission of vengeance are: "The time is out of joint. O cursed spite, / That ever I was born to set it right."[7] Time and place, the ineluctable modalities of action, inevitably restrain as well as enliven the heroic man's limitless drive for self-expression. Hence Hamlet's and Lawrence's alternate chafing at and championing of the occasion for such self-expression to which action so perversely calls them. Both finally find a way through their untimeliness to some mastery of idealizing action's limitations, Hamlet through a Christian acquiescence to them, Lawrence through substitution of the ancient triumphal form for the epic one his action seemingly demanded.

What value in life, and in his, which was forged in loving Dahoum, in fighting to give this Arab and his brethren freedom, and in writing that tale, did Lawrence's inventive literary switch allow him to express? A soldierly one, ironically enough, namely courage—specifically the courage to affirm life itself, and one's own, when both come up short. But not a conventionally soldierly one, not legendary "epic" courage, which can make no room for that craving necessarily disappointed in any action. Lawrence could not feel "such men as Auda" in himself because he understood that Bedouin chieftain as a storybook, adventuring hero. Fair enough. Idealized action in pursuit of an absolute should be allowed some scope for fantasy. That Lawrence takes that liberty and pays the cost by acknowledging how much inadequacy accompanies any enactment of the absolute, how much humbling accompanies his erotic and military "triumph" in the Arab Revolt, is the true measure of his glory in the anti-epic *Seven Pillars of Wisdom*.

Reporting the inadequacy, the unanswered craving that compromises satisfaction in one's own achievements, always runs the risk of appearing grandiose, even narcissistic. Lawrence has certainly been rewarded with such opprobrium by his detractors from the start. But his casting of the report as a "triumph" and his persistent magnification of that form's humbling of the victor are poignant, magnanimous enough defense against such scorn. The freedom Lawrence finally achieved in his narrative is not so much self-expression as it is courageous telling of the truth about action: that it calls one to imperfection, to incompleteness, to the ultimate, permanent encounter with unappeasable craving. Such is the ironically biblical wisdom Lawrence's erotic, literary house provides.

Lawrence makes much of the parallels between himself and Jesus in *Seven Pillars of Wisdom,* normally to denigrate the authenticity of his sacrificial engagement with the Arab cause. There is a generous parallel to draw as well, offered by the absolute which governs this tale of his craving, namely, freedom. In John 8:13–14 the Pharisees reject Jesus' authority to speak of the value of life as follows: "'Thou bearest record of thyself; thy record is not true.' Jesus answered and said unto them, 'Though I bear record of myself, *yet* my record is true: for I know whence I came, and wither I go.'" Later in the same chapter, speaking in verse 32 to an audience willing to believe him, Jesus finishes his report in a way that Lawrence would not claim for himself but that bears on his tale's accomplishment, at least for its readers, if not entirely for him: "And ye shall know the truth, and the truth shall make you free."

Notes

1. Simon Hornblower and Anthony Spawforth, eds., *The Oxford Classical Dictionary: The Ultimate Reference Work on the Classical World* (Oxford and New York: Oxford University Press, 1996), 1554.
2. T. E. Lawrence, *Seven Pillars of Wisdom* (Garden City, New York: Doubleday, Doran & Company, 1935), 29.
3. Ibid., 5.
4. Ibid., 5.
5. Ibid., 549. For an extended discussion of epic, history, and incompleteness in *Seven Pillars of Wisdom,* readers should consult Charles M. Stang's essay, "Does Not Care," in this volume.
6. Ibid., 549.
7. William Shakespeare, *Hamlet* (London and New York: Methuen, 1982), 228.

Chapter Two

"Does Not Care"

Lawrence, Herodotus, and Nietzsche on History

Charles M. Stang

> I shun father and mother and wife and brother when my genius calls me. I would write on the lintels of the door-post, *Whim*. I hope it is somewhat better than whim at last, but we cannot spend the day in explanation.
>
> —Ralph Waldo Emerson, *Self-Reliance*

In 1924, T. E. Lawrence had a stone lintel placed over the door to Clouds Hill, the remote cottage in Dorsetshire where, until his death in 1935, he occasionally lived and often entertained friends and family. Into this stone lintel Lawrence carved two Greek words: ου φροντις, "does not care." He explained to friends that these words were taken from a comic episode in the *Histories* of Herodotus, one concerning a young suitor Hippoclides and a rich king Clisthenes, the father of the young woman whose hand Hippoclides was seeking in marriage. Hippoclides was one of many young noblemen vying for the hand of the king's daughter. Wishing to give his daughter to the very best man in Greece, Clisthenes devised contests for the suitors. In all of these contests young Hippoclides excelled. When the day came for Clisthenes to announce his choice, however, there was a final feast with abundant song and drink. Hippoclides, quite drunk, bade the musicians strike up a dance. And just as in the other contests, Hippoclides excelled, dancing beautifully. But the drink in him spurred him on, and he was soon dancing on the tables. When at last Hippoclides stood on his head upon the table, tossing his legs about in the air and exposing himself to the whole feasting party, Clisthenes, furious at this shameful behavior, cried out, "Son of Tisander, you have danced your marriage away!" "Hippoclides does not care," the young reveler replied.[1]

Between the years of 1924 and 1935, Lawrence would often present himself as a modern Hippoclides, freed from all care. The first evidence of this new self-identification comes from a letter to A. E. "Jock" Chambers, a friend from the Tank Corps, dated August 8, 1924. Lawrence is writing to Jock from Clouds Hill nearly a year before he added the Greek inscription to the lintel: "This address is my safest one: it may be any name. 'Shaw' I call myself, but some Ross and others Lawrence. Hippoclides doesn't care!"[2] "Ross" and "Shaw" are the two names Lawrence assumed after the war in order to secretly enlist in the ranks of the RAF and Tank Corps and also to attempt, however vainly, to avoid the attention of the press.[3] Alec Dixon, another friend from the Tank Corps, reports that on an afternoon in July 1925: "I arrived at Clouds Hill to find its owner at the top of a step-ladder busily cutting an inscription on the concrete lintel of the doorway. . . . Shaw smiled down at me, and stopped work to tell the story of Hippoclides. . . . [Shaw said of ου φροντις] 'Don't give a damn,' is a good translation, I think."[4] When Lawrence's biographer Robert Graves published a letter from 1927 in which Lawrence again presents himself as Hippoclides, here with regard to publishing details, Graves felt the need to provide the reader an explanation in a footnote. Graves calls Hippoclides "one of Lawrence's odd self-identifications" whose "not caring is willful illogicality."[5] Finally, in a letter to Celandine Kennington dated October 18, 1932, Lawrence provides what is perhaps the most robust explanation of this self-identification:

> Dear Celandine, In Athens was a gentleman called Hippoclides who became engaged to a rich merchant's daughter: and they arranged him a slap-up and splendid marriage. The feast preceding it was too much for his poor head, though. He stood on his head on the table and did a leg-dance, which was objectionable in Greek dress. "Hippoclides, Hippoclides" protested the shocked merchant "You dance your marriage off." "Wyworri?" said Hippocleides [*sic*]: and Herodotus tells the tale so beautifully that I put the jape on the architrave. It means that nothing in Clouds Hill is to be a care upon its habitant. While I have it there shall be nothing exquisite or unique in it. Nothing to anchor me.[6]

Around this simple Greek inscription revolves a constellation of suggestive meanings. The Greek reads ου φροντις Ιπποκλειδη, literally "Hippoclides has not a care." The noun Φροντις suggests the thought, care, or attention that one bestows on a person or a thing.[7] But here on the lips of Hippoclides and on the lintel of Lawrence there is no object expressed, and so the noun takes on the sense of a general condition: reflection and meditation, care understood as the concernful attachment to the world, even anxiety. All of these conditions are negated by the ου, leaving a sense of anything from whimsical dismissal to blissful detachment to apophatic unthinking. Add to this complex two other curious valences. Φροντις may also mean "heart's desire"; its denial suggests that the heart can no longer venture to desire or that the object of the heart's desire is gone. This begs comparison with the famous dedicatory poem of *Seven Pillars of Wisdom,* where Lawrence addresses his beloved "S. A.," a young Arab man named "Dahoum" who died just before the end of the revolt.[8] Φροντις may also mean authority, office, function, or department. While Hippoclides forfeits his marriage, Lawrence forswears all position: his rank of colonel exchanged for a life in the ranks, the limelight for anonymity, and "Lawrence" for "Ross" and "Shaw."

At the center of this whirl of meanings we sense Lawrence's exhausted renunciation of the world's elevated tasks and rewards, the renunciation that would be for him both plague and palliative in the years following the war. In the conflict between young Hippoclides and Clisthenes over an "objectionable" dance, we hear echoes of a ruined revolt, an immodest venture of youth wrecked by old men: "When we achieved and the new world dawned, the old men came out again and took our victory to re-make in the likeness of the former world they knew. Youth could win, but had not learned to keep: and was pitiably weak against age. We stammered that we had worked for a new heaven and a new earth, and they thanked us kindly and made their peace."[9] In the insistence that there be nothing "unique or exquisite" in Clouds Hill, we hear Lawrence's aspirations to anonymity, the prayers of a "great man" to be forgotten, to be made instead absolutely average, indeed at times even abased. In the hope that nothing is to be a care, we hear the voice of a certain Galilean promising that one can in this world be like a lily in the field. In the hope that nothing is to anchor, we hear this same voice again, warning against our laying up treasures where thieves come near and moths destroy. Perhaps these uncaring aspirations seem illogical, as they do to Graves, but they are not willful. Or if willful, they are paradoxically so: those willful exertions, known well to all ascetics and mystics, by which the aspirant annihilates his will.

But Lawrence was ever the prankster. To supporters and detractors alike, friends and strangers, even to his own biographers, Lawrence presented himself and his purposes ambiguously. Lawrence was often playful and always learned. We would do well, then, to ask whether there might be more play and learning in this simple Greek inscription than is first apparent. What more might ου φροντις suggest? Although he always deemed himself inferior to the literary giants of his day, Lawrence does occasionally manage to achieve Eliot's allusiveness and Joyce's playfulness. And although Lawrence seems to agree with Emerson that "we cannot spend the day in explanation," we who follow can—and will.

Herodotus Does Not Care

There is another infamous instance of ου φροντις in the literature of antiquity. In his uncharacteristic polemic *On the Malice of Herodotus,* the philosopher-historian Plutarch says of the Father of History: "It looks to me as though, like Hippoclides standing on his head on the table and waving his legs in the air, Herodotus would 'dance away the truth' and say: '*Herodotus doesn't care*'"(my emphasis).[10] Plutarch uses this comic episode from the *Histories* to suggest that Herodotus himself is a sort of Hippoclides, that his entire historical project is best summed up by the phrase "does not care." This signals not so much Herodotus' willful illogicality as it does his moral failure: κακοηθεια. It is difficult to find a word in English that captures the sense of this charge: "malice" is often used, sometimes "malignity." The word suggests a moral offense, a viciousness of character prone to put the worst construction on everything, and outrageously bad manners and habits.[11]

For Plutarch, that Herodotus "does not care" means that he has abandoned his commitment to truth in favor of a whimsical license. He struggles but eventually fails to find in Herodotus' narrative a concerted intelligence behind these departures from truth. Sometimes for reasons stylistic, sometimes for reasons political, sometimes for reasons unknown, Herodotus stands on his head and dances. Plutarch is forced to attribute these inconsistencies not to a hidden agenda, but to a flaw of character.

Let us put aside for the moment the particulars of Plutarch's critique of Herodotus. More important for our investigation is the fact that Plutarch identifies Herodotus himself with Hippoclides by way of the phrase ου φροντις. This provides us a critical lead for understanding Lawrence: what if we, following Plutarch, were to understand the inscription above Lawrence's door as his identification not only with Hippoclides but with Herodotus himself? What if we were to understand Lawrence as aligning himself with Herodotus, writing history of the sort that Herodotus did and thereby also opening himself to Plutarch's criticisms? What if we were to situate Lawrence thus, in an ancient debate over how one should care and how one should write—more specifically, how one should care for and write what has come to be called "history"? What might we learn of how Lawrence understood writing and how we consequently might read him?

Herodotus is one among a small group of classical authors (including Homer, Aristophanes, Thucydides, Xenophon, and Plato) to whom Lawrence makes reference again and again in his writings, especially in his correspondence. In 1930, his old friend David Garnett invited Lawrence to edit a new edition of Herodotus' *Histories* for the Nonesuch Press. In a letter dated February 14 of that same year, Lawrence declines the invitation but writes at some length about Herodotus:

> I once knew Rawlinson's *Herodotus* well. It is—respectable, I think. Not as good as Herodotus, one of the best of men, deserves: but what he got . . . Herodotus is a marvellous fellow. I incline to credit everything he saw, and to give him credit for intelligence and judgment in sifting the stories that he heard. Also he is pre-eminently human. A first-rate *Herodotus* would be a most useful book: only I'm worried to think of Rawlinson as the vehicle. Good of course, very good, as scholarship goes: but Rawlinson was too sober.[12]

Evident here is the great feeling Lawrence has for Herodotus; the historian is "one of the best of men," a "marvellous fellow," "pre-eminently human." One may suppose that Lawrence feels some solidarity with Herodotus, who received "not as good as [he] . . . deserves: but what he got." What is most interesting in this letter, however, is the fact that Lawrence believes George Rawlinson's translation of Herodotus to be inadequate because he was "too sober." Lawrence understands that an appreciation of Herodotus requires an appreciation of his playfulness, his curious mixture of irony and whimsy. And this brings us back to the figure of the drunken Hippoclides and the too sober Clisthenes. To Plutarch's critical suggestion that Herodotus is no better than Hippoclides, Lawrence answers that he is no worse, that the two are both playful in a way that establishes them as the "best of men" and "pre-eminently human." It is to their rank that he aspires when he carves over his door their playful proverb.

But what of Herodotus and Lawrence, what of the *Histories* and *Seven Pillars of Wisdom?* Can they be plausibly compared? E. M. Forster, one of Lawrence's many literary friends, writes of *Seven Pillars of Wisdom:* "Round the tent-pole of a military chronicle, T. E. has hung an unexampled fabric of portraits, descriptions, philosophies, emotions, adventures, dreams. He has brought to his task a fastidious scholarship, an impeccable memory, a style nicely woven of Oxfordisms and Doughty, an eye unparalleled . . . a profound distrust of himself, a still profounder faith."[13] Excepting the comments on Lawrence's style, this could just as easily describe the achievement of Herodotus. Both narratives have at their center a war, but that center is often obscured by digressions and digressions upon digressions. As Herodotus admits in his fourth book, "From the start my narrative has quite naturally led to digressions."[14] These digressions include character sketches, detailed geographical descriptions, stirring speeches, philosophical meditations, and fantastic tales. Some scholars believe that Herodotus began his history as a simple account of his travels but thereafter realized how he might organize the whole around the rise of the Persian Empire and its defeat at the hands of the Greeks.[15] Before the outbreak of World War I, while he was still an archaeologist in Syria, Lawrence intended to write a book entitled *Seven Pillars of Wisdom,* an account of his travels through seven cities of the Near East. Later, when he thought to compose an account of the Arab Revolt, he kept that same title (now with new resonances) and managed, much like Herodotus, to incorporate into the narrative much of the material from his previous wanderings.[16]

It is not uncommon to find Books 7 through 9 of Herodotus' *Histories* published independently under the name *The Persian Wars.* In these last three books, the narrative of the events of the Persian Wars is less obscured by digressions and offers a cleaner chronicle to the reader less interested in ethnography or etiology. Similarly, Lawrence published an abridged version of *Seven Pillars of Wisdom* entitled *Revolt in the Desert.*[17] Here is the story of the revolt without Lawrence's detailed landscapes, ethnographic studies, or tortured confessions. Lost in these abridgments is the distinctive achievement of these two artists: an epic history that at times drops its carnival mask to reveal a beautiful mosaic.

Several clues in the *Histories* suggest that Herodotus left his epic unfinished. The promise of an account of Assyria goes unfulfilled, as does the promise to tell us of the fate of the infamous traitor Ephialtes. In the dedicatory poem, Lawrence announces the same regarding *Seven Pillars of Wisdom:*

> Men prayed me that I set our work, the inviolate house,
>   as memory of you.
> But for fit monument I shattered it, unfinished."[18]

There is a rich ambiguity as to what "our work, the inviolate house" is and what "fit monument" Lawrence leaves as a memory of his dead beloved. Does "the inviolate house" refer to the Arab Revolt itself, the aim of which was "to build an inspired dream-palace"?[19] Or does it refer to the book in hand, whose title makes reference to another house from Proverbs 9:1, "Wisdom has built her house, she has hewn her seven pillars"? If indeed Herodotus' *Histories* are unfinished, this may be to suggest that the panoply of causes in history can never be exhaustively set down, that our questions always overreach our answers, that however far we cast the net of inquiry—and how far Herodotus casts that net!—there will always be an Assyria or an Ephialtes whose part in the great drama is lost. And if indeed Lawrence's *Seven Pillars of Wisdom* is left shattered, unfinished, this may be to suggest not only that written history must reflect the events which it sets out to record, that a narrative of a ruined revolt must be left something of a ruin, but that the historian himself must also be marked by the tragedy and triumph he records, that a man who is witness to men's efforts crashing against necessity must himself be left shattered, and that a man addressing a dead beloved cannot speak as one fully alive.

Finally, both narratives begin with a poem announcing something of the aim and scope of the history that is to follow. The "proem" of Herodotus reads thus: "These are the researches of Herodotus of Halicarnassus, which he publishes, in the hopes of thereby preserving from decay the remembrance of what men have done, and of preventing the great and wonderful actions of the Greeks and the Barbarians from losing their due meed of glory; and withal to put on record what were their grounds of feud."[20] And the dedicatory poem of *Seven Pillars of Wisdom* begins with these oft-quoted lines:

> I loved you, so I drew these tides of men into my hands
> and wrote my will across the sky in stars
> To earn you Freedom, the seven-pillared worthy house,
> that your eyes might be shining for me
> When we came.

When Lawrence did come north with the Arab Revolt, "S. A." was already dead: "The strongest motive throughout had been a personal one, not mentioned here, but present to me, I think, every hour of these two years. Active pains and joys might fling up, like towers, among my days: but, refluent as air, this hidden urge re-formed, to be the persisting element of life, till near the end. It was dead before we reached Damascus."[21] Shattered work, an unfinished history—this is the "fit monument" Lawrence leaves his dead beloved, "a memory of you."

But how does memory figure for these two historians in their opening poems? Herodotus wants to preserve for posterity the noble deeds of Greeks and barbarians. The "due meed of glory" is not the exultation of fellow warriors after victory, or even the triumphant parade through the city. Such heroes as defeated the Persians deserve undying glory. Memory struggles against ceaseless becoming, against the ebb and flow of coming-to-be and passing-away. Glory, as headstrong as its heroes, demands that it be exempted from this metaphysical necessity, that it be granted a portion of unchanging being.[22]

In his Introductory Chapter, Lawrence writes: "It seemed to me historically needful to reproduce the tale, as perhaps no one but myself in Feisal's army had thought of writing down at the time what we felt, what we hoped, what we tried." It is significant that what Lawrence struggles to rescue from oblivion is not so much the deeds of his Greeks and barbarians, his British and Arabs, but their emotions, their aspirations, their efforts. Here Lawrence is something of a *modern* historian, not in the sense of writing a more scientific history but in the sense of trying to write a history of interiority. Also modern, more specifically romantic, is the fact that Lawrence has a different relationship to decay than does Herodotus. For Herodotus, decay is the enemy and his history is an attempt, however vain, to rescue men's deeds from oblivion: "in the hope of thereby preserving from decay the remembrance of what men have done."[23] Lawrence cannot figure decay as the enemy, for he imagines his beloved, the wellspring of his own glorious actions, in a state of decomposition:

> Death seemed my servant on the road, till we were near
> and saw you waiting:
> When you smiled, and in sorrowful envy he outran me
> and took you apart:
> Into his quietness.
>
> Love, the way-weary, groped to your body, our brief wage
> ours for the moment
> Before earth's soft hand explored your shape, and the blind
> worms grew fat upon
> Your substance.[24]

His love for this dead man compels him to give something of himself and something of his history over to decay. Neither he, nor his beloved, nor even his history will escape decomposition. Perhaps his history is not the bulwark of being against becoming that Herodotus' aspires to be. Perhaps Lawrence leaves his history unfinished so that:

> The little things creep out to patch themselves hovels
> in the marred shadow
> Of your gift.[25]

That his history is left so open to decay—indeed that he acknowledges decomposition at the inaugural moment of composition—may be to suggest the vanity of history's aspiration to wrest being from becoming, the folly in supposing that history is about the business of preservation over against decay, memory over against oblivion.

On the Malice of Herodotus

Apart from the inscription over the door, we have evidence from his letters that Lawrence greatly admired Herodotus. And apart from Plutarch's suggestion, we have evidence from his letters that Lawrence understood Herodotus to be playful, much like the drunken Hippoclides. We have surmised from this evidence that just as Lawrence identifies himself with Hippoclides, he identifies himself with Herodotus. We have fruitfully compared Herodotus' *Histories* with Lawrence's *Seven Pillars of Wisdom,* specifically over the questions of structure, genesis, abridgments, completion, and openings. This pairing of Herodotus and Lawrence, however, will become even more compelling when we consider whether and how Plutarch's criticisms of Herodotus might also apply to Lawrence. What is most interesting in these criticisms is not that they discredit the historical projects of Herodotus and Lawrence, but rather that they help explain the aim and scope of those projects.

Plutarch begins his rather shrill essay with the charge that many readers have been deluded by the style of Herodotus, deluded into believing that he records the truth and that he is of good moral character. Plutarch claims to stand for the cause of truth and for the cause of his ancestors (the Boeotians, whom Herodotus claims were treacherous to the rest of the Greeks in the Persian Wars) when he declares that everywhere Herodotus is motivated by malice. He lays before the reader seven indications by which one can determine whether a narrative is written with malice or honest integrity: (1) whether the narrator uses severe language when gentler will do; (2) whether he includes discreditable information irrelevant to the immediate narrative; (3) whether he withholds due praise; (4) whether he prefers the less creditable version of events; (5) whether he attributes glorious successes not to valor but to money, little resistance, or good luck; (6) whether he slanders indirectly; (7) whether he damns with faint praise. We are not surprised to discover that Plutarch finds Herodotus guilty of each of these infractions. Although Herodotus has had his many critics, ancient and modern, none has come close to the vitriol of Plutarch. Even modern critical readers of Herodotus are surprised by this essay. Indeed some scholars, embarrassed that Plutarch would have written such a shrill polemic, argue that this essay is not the authentic work of the otherwise pacific philosopher-historian, but rather that of a disgruntled follower.[26]

Chief among Plutarch's complaints against Herodotus is his account of the causes of the Trojan War. Herodotus tells of a series of kidnappings of prominent women between the East and the West. When Alexander (Paris in other versions), son of Priam, carried off Helen, he was simply answering a former kidnapping by the Greeks. The Greeks, however, did not acknowledge the parity and so raised an army to destroy Troy and reclaim Helen. Plutarch objects that Herodotus has thereby made the Trojan War, "the greatest and noblest exploit of Hellas," little more than "an act of folly, entered upon for the sake of a worthless woman."[27] Scandalized, Plutarch asks, "Well, then, is there anything glorious or great left to the Greeks from these battles?"[28] Against Herodotus' various and inglorious causes of the Trojan

War, Plutarch fashions himself the defender of the Homeric tradition, the defender of the glorious memory of Hellas.

Perhaps more appalling than his challenging the "greatest and noblest exploit of Hellas" is the fact that Herodotus reveals himself to be a φιλοβαρβαρος, a "barbarian-lover."[29] This fascination with the Eastern Other, Plutarch argues, pollutes Herodotus' entire narrative. Against the myth that the Egyptians had once attempted to sacrifice Heracles to Zeus, Herodotus answers: "such a story proves the Greeks to be utterly ignorant of the character and customs of the [Egyptian] people"—indeed, the Egyptians will not even sacrifice beasts, never mind humans.[30] Herodotus believes a story told him by the Egyptians that Menelaus, like his brother Agamemnon, performed human sacrifice in order to win better sailing weather.[31] He claims that the Persians learned pederasty from the Greeks, while the Greeks learned of their twelve gods from the Egyptians.[32] It is not the barbarians who sacrifice humans, but the Greeks; it is not the barbarians who teach the Greeks debased practices, but the Greeks who learn of divine things from the barbarians. One doubts whether this barbarian-loving prejudice is as dominant in the *Histories* as Plutarch presents it. More often than not, Herodotus presents Greece as a blessed order amidst barbarian chaos. Regardless, Plutarch is convinced that this prejudice amounts to a betrayal of the traditions of Hellas, that Herodotus is guilty of malice, impiety, and blasphemy.[33]

Plutarch concludes his criticisms thus: "We must admit that Herodotus is an artist, that his history makes good reading, that there is a charm and skill and grace in his narrative, and that he has told his story 'as a bard tells a tale,' I mean not 'with knowledge and wisdom,' but 'with musical clear-flowing words.'"[34] The name Plutarch gives Herodotus, here translated "artist," is γραφικος, meaning someone suited for and skilled in writing, drawing, or painting. The sense is of someone whose mimetic enterprise, be it on the page or the canvas, is governed more by the concerns for lively description than by the strictures of veracity. In this passage, Plutarch goes on to liken Herodotus to none other than Odysseus. And while earlier Plutarch fashions himself the defender of the Homeric tradition, this fruitful pairing of Herodotus the wandering inquirer with the crafty hero of the *Odyssey* is not meant as a compliment to the artist-historian. Plutarch deliberately misquotes King Alcinous, who, after hearing Odysseus sing of his amazing wanderings, says to him: "In your words is a formal beauty to match the graceful order of your ideas: you frame and bedeck this tale of the Argives' hardships and your toils as knowingly as any bard."[35] For Plutarch, that for which Alcinous (perhaps ironically) praises Odysseus is exactly that for which Herodotus should be blamed: he is a bard masquerading as a proper historian whose ignorant tale may sound sweet but hides his "slanders and ugly lies."

On the Malice of Lawrence

Like Herodotus, Lawrence has had many detractors. And like Herodotus, he has been charged with malice, with the seven offenses against honest history. Richard Aldington, the most vitriolic of Lawrence's British critics, claims to have discovered "a systematic falsification and over-valuing of himself and his achievements which Lawrence practised from a very early date. In other words the national hero turned out to be at least half a fraud."[36] Systematic falsification and fraud—these sound very much like Plutarch's general criticism of Herodotus. But let us take a more specific example from Plutarch. He claims that Herodotus is guilty of attributing the glorious successes of the Greeks to, among other ignoble factors, the influence of money. He has in mind here Herodotus' account of Themistocles, the great Athenian hero of the Persian Wars, who he claims only reluctantly and only with a substantial bribe agreed to have his fleet meet the Persians at the battle of Artemisium. Similarly, Lawrence suggests that Auda Abu Tayi, the only Homeric hero in *Seven Pillars of Wisdom,* nearly betrayed the other Arabs and their revolt when the Turks offered him more money than the British.[37] Not surprisingly, objections to such details as this have come not from British but from Arab critics, most notably Suleiman Mousa, who feels that Lawrence, by exaggerating his own role in the revolt, has robbed the Arabs of the glorious success due them: "one seeks in vain for a semblance of justice to the Arabs."[38] Still controversial is the question of how substantial a role Lawrence played in the revolt. Less controversial is the claim that the Arab Revolt, whoever was at its helm, was not the glorious success that Lawrence once hoped it would be and Mousa longs for it to have been.

War: Glorious Exploit or Great Folly?

Unlike some of his postwar intellectual friends, such as Siegfried Sassoon, Lawrence never made a great noise about the inglorious causes of World War I. He did, however, walk what seemed to many a dangerous line between hailing the Arab Revolt as a noble and glorious exploit on the one hand, and condemning it as a grave folly on the other. His position as to which it was depended in part on whom he was addressing, whether it was an international delegation of diplomats or a trusted friend. But his oscillation was not only cold calculation: he was himself torn over how to understand the events in which he took part. *Seven Pillars of Wisdom* stands as a testament to this tortuous struggle, a first draft of which was written during the Paris Peace Conference in 1919, when Lawrence lobbied for the Arabs against French colonial interests, and subsequent drafts of which were composed in the early 1920s, when Lawrence was working with Winston Churchill, busily redrawing the map of the Middle East to serve British interests. Lawrence's deep ambivalence is given clearest expression in the Introductory Chapter, where he writes, with one hand: "I meant to make a new nation, to restore a lost influence, to give twenty millions of Semites the foundations on which to build an inspired dream-palace of their national thoughts"[39] and, with the other hand: "[The Arabs] grew accustomed to believing me and to think my Government, like myself, sincere. In this hope they performed some fine things, but, of course, instead of being proud together, I was continually and bitterly ashamed."[40]

Lawrence's increasing shame over his role in the Arab Revolt made him very unpopular among those in the government who wished to use the revolt for their own purposes. A seemingly perfect distraction from the horrors of trench warfare in France, the Arab Revolt presented the disenchanted public with images of an ideal war: romantic, noble, successful. "Lawrence of Arabia" was the hero of this Arabian adventure, whether T. E. Lawrence wished it or not. That Lawrence did not sufficiently endorse this presentation of the revolt and that he often undermined it made him something of a pariah in certain government circles. When we read the Introductory Chapter of *Seven Pillars of Wisdom* and soon understand the bitter irony of the subtitle "A Triumph," we can easily imagine a British Plutarch, fed on the patriotic glories of his own Hellas, asking, "Well, then, is there anything glorious or great left to the Greeks from these battles?"[41] Has Lawrence not made us wonder whether even this Trojan War, the Arab Revolt, was, far from "the greatest and noblest exploit of Hellas," rather "an act of folly"?

Lawrence the Barbarian-Lover

Lawrence's presentation of the Bedouin Arab has drawn criticism from two camps. The first camp consists of those who feel that Lawrence unjustly glorifies the primitive Bedouin over the civilized Englishman and that his prejudice for the desert over the sown pollutes his narrative. Such critics, who were his British contemporaries, would feel at home with Plutarch, and their equivalent of the charge that Herodotus loved barbarians would be that Lawrence "went native." Throughout *Seven Pillars of Wisdom,* Lawrence introduces the reader to other British soldiers and officers, from the ranks of the Camel Corps to the General Headquarters in Cairo, who do not share his appreciation of the Arabs. Like Herodotus, Lawrence was committed to schooling his countrymen in the ways of the Eastern Other. And like Herodotus, Lawrence was convinced that the success or failure of his nation's military endeavor depended on this schooling. Before he was the liaison officer to Feisal in the Hejaz, Lawrence wrote for his fellow British officers a short manual entitled "Twenty-Seven Articles," in which he attempted to show the value of Arab, especially Bedouin, customs and the value of British acclimation to those customs in the field.[42] *Seven Pillars of Wisdom* and the other postwar writings can be read as a sort of apology for the Arabs, a glorifying account of their cultural strengths that aspires to explain both why the British helped them in their revolt and why the British should now support their nationalist aspirations against French colonial interests.

But Lawrence's presentation of the figure of the Arab has more recently drawn criticism from another camp. In his cultural polemic *Orientalism,* Edward Said has placed Lawrence in a long tradition of Western scholars and adventurers who have reduced the panoply of Arab culture to a single reified figure: the Arab.[43] Sometimes these orientalists present a few such figures, as Lawrence does: desert Arab and settled Arab, Bedouin, and Levantine. This ideological construct, the Arab, acts as a foil to the West's understanding of itself, sometimes offering a decadent Other that highlights the achievements of occidental civilization, sometimes offering a pure and noble savage for which an alienated West can pine. Not surprisingly, these criticisms cannot be easily associated with Plutarch's, or any other ancient critic's, charges against Herodotus. We would be hard-pressed to find the ancients criticizing their historians and ethnographers on the grounds that they present an exotic Other that effaces the particularities of the barbarian. While Said and those who have written in his wake have articulated some powerful criticisms, they have also often ignored or effaced those moments in which the orientalists themselves, including Lawrence, seem to understand and resist the tradition of which they are part. But as is evidenced in Plutarch's essay on Herodotus, polemics always require selective memory, and the orientalist critique is no exception.

Odyssean Bard Weaving an Artistic Tale

Finally, what of Plutarch's charge that Herodotus is an artist, a spinner of yarns much like Odysseus? How is Lawrence open to this same charge? While other essays in this volume examine the relationship of Lawrence to Odysseus in some detail, we would do well to consider here whether and how Lawrence, like Herodotus, can be understood as an artist, a γραφικος. In a letter to his old friend Vyvyan Richards in early 1923, Lawrence addresses point by point Richards's responses to *Seven Pillars of Wisdom* (the underlined phrase and sentence are presumably Lawrence's paraphrases from Richards's earlier letter):

> More critic than artist. That's the analytic vein in me. Ineradicable. A critic in conscious creation is of course an artist. . . . A critic is no more barred from creation than any other human being. . . . [T]he perfect artist is half-critic & half-creator.

> Written history has never yet so nearly approached the unity of a work of art. Written history is long & must be judged by the standard of epic rather than lyric. . . . I suppose the *Peloponnesian War* has more unity than the average drama. Only perhaps you'd call it not true. I'll admit that modern history has seldom been "composed" in the artistic sense. . . . But modern history tries to be a science, not an art.[44]

Not satisfied with Richards's antinomy between creation and criticism, Lawrence reclaims for himself the title "artist." He confesses that he will always be a critic, but argues that "the perfect artist is half-critic & half-creator." Lawrence certainly does not think himself "the perfect artist," but an artist nonetheless. He enthusiastically accepts the title that Plutarch gives Herodotus as a criticism. His art takes the form of written history, which under Lawrence's pen, Richards suggests, "has never yet so nearly approached the unity of a work of art." Despite his earlier suggestion that Lawrence is more critic than creator, Richards never doubts that his old Oxford chum is indeed a first-rate artist.

As has been mentioned, there is in the title γραφικος not only the sense of one skilled in writing, but also one skilled in drawing or painting. This suggests that writing is aligned with the visual arts. The 1926 subscribers' edition of *Seven Pillars of Wisdom* is filled with visual art: painted and drawn portraits, landscapes, photographs, even cartoons. Lawrence meant these images to communicate with the written word. One of the more curious examples is a series of Lawrence's own photographs of the streets of Jidda, a small city on the coast of the Red Sea. In a letter to Charlotte Shaw in 1927, Lawrence mentions these images and their relation to the writing they accompany: "My views [in *Seven Pillars of Wisdom*] are like my photographs of Jidda: the edges, even the sharpest, are just modulated off, so that you can't put a pin point on them. Drawn, not in line, but in tone. Atmospheric. It's the difference between impressionism and the classical."[45] More important for our purposes than the character of the relationship between impressionism and the classical is the fact that Lawrence testifies to the strong connection between the artistry of his writing and the artistry of his photography. Lawrence is truly a γραφικος in all the word's many senses.

Lawrence explains that his art, written history, must be judged by the standard of epic. In his short but compelling treatment of the Greek historians, T. J. Luce explains how Greek history grew out of the epic tradition, especially the Homeric epics.[46] In his *Histories,* Herodotus indeed manages to weave together Homer's two great *subject matters:* a narrative of an all-important war, as in the *Iliad,* with a detailed description of foreign lands and peoples, as in the *Odyssey.* The *form* of Greek history also borrows from Homer: a narrative of events, detailed yet circumstantial, including most often what persons (including gods in Homer) did and said. Persons are the focus—individual actors, not institutions (military, religious, or political), social classes, economic forces, or cultural achievements. By his insisting that his history be judged by the epic standard, Lawrence is aligning himself with the historical tradition that emerged from the epics, a tradition wherein Herodotus is king.

He goes on to speak of "modern history" as something alien to his enterprise. While Lawrence is an artist and his history art, modern history "has seldom been 'composed' in the artistic sense." But artistry in composition was central to the early Greek historians: "History was therefore a literary enterprise above all, the imaginative creation of its author. It was to be written in the language of its creator . . . History was the creation of the historian, not a mirror he held up to a subject that had an objective and independent existence apart from himself and which he might transfer faithfully to the page."[47] Lawrence playfully underscores Luce's point in his essay "Evolution of a Revolt," when he writes that "the prejudices of historians are generally the richest part of their histories."[48] The understanding of history as a mirror held up to an objective reality is what Lawrence terms "modern": "modern history tries to be a science, not an art." Such a scientific history adheres to what Lawrence considers a very narrow understanding of truth, truth as objective and verifiable. According to these standards, history of the sort he is composing is found wanting: "Only perhaps you'd find it not true."[49] Again Luce supports Lawrence's perspective: "From what has been said it should be clear that the classical historians had a different conception of historical truth than we do today. . . . [T]here was from our modern viewpoint a great deal of 'untruth' in ancient historiography." Lawrence announces this fact in the concluding sentence of his Introductory Chapter: "In this book also, for the last time, I mean to be my own judge of what to say," and again in a letter to his friend H. S. Ede, "the interests of truth and form differ."[50] And his first sensational biographer, Lowell Thomas, reports that Lawrence once declared, in true Hippoclides fashion: "History isn't made up of truth anyhow, so why worry?"[51]

The word Plutarch uses in his concluding criticisms to describe Herodotus' writing is not ιστορια, literally "inquiries," often translated "histories."[52] Instead, never missing an opportunity for a last snub, Plutarch calls Herodotus' *Histories* a λογος, here meaning a "story." With this same word Lawrence begins *Seven Pillars of Wisdom:* "The *story* which follows . . ." (my emphasis).[53] Again, at the opening of the next chapter, Lawrence describes his literary enterprise: "Some of the evil of *my tale* may have been inherent in our circumstances" (my emphasis again).[54] A story and a tale—does this mean that *Seven Pillars of Wisdom* is not history? So Lawrence seems to suggest in his Introductory Chapter: "In these pages the history is not of the Arab movement, but of me in it. It is a narrative of daily life, mean happenings, little people. Here are no lessons for the world, no disclosures to shock peoples. It is filled with trivial things, partly that no one mistake for history the bones from which some day a man may make history, and partly for the pleasure it gave me to recall the fellowship of the revolt."[55] And yet Lawrence, in his letter to Vyvyan Richards and in countless other instances, refers to *Seven Pillars of Wisdom* as history. Which is it then: a story or a history? The question is poorly posed, or rather presumes a certain understanding of history as opposed to story. When Lawrence here distances himself from history, he is again speaking of the modern history which tries to be a science. From the perspective of this modern scientific history, his own history amounts to nothing more than "bones," at best an incomplete and questionable source for the reconstruction of the events of the Arab Revolt. Are Herodotus' *Histories* story or history? Is Lawrence's *Seven Pillars of Wisdom* story or history? The critic Richard Aldington offers an answer: "Whatever its literary merit (which may be very high) *Seven Pillars of Wisdom* is rather a work of quasi-fiction than of history. . . . If *Seven Pillars of Wisdom* is to be regarded as a work of the imagination rather than of history, it may well be considered by its admirers all the more remarkably for that fact."[56] As is clear from Luce's account of Greek history and Lawrence's discussion of his own literary enterprise, Aldington is quite wrong. There is a sort of history that cannot be clearly distinguished from story, there is a sort of historian who cannot be clearly distinguished from Odysseus.

Lest we suppose that Herodotus' *Histories* and Lawrence's *Seven Pillars of Wisdom* are works of "quasi-fiction," we would do well to remember that much of what is reported in both narratives is indeed true in the sense of objective and verifiable. There is no better source for understanding the course of the Persian Wars than Herodotus or the course of the Arab Revolt than Lawrence. Herodotus' narrative of the rise of the Persian Empire and its defeat at the hands of the Greeks has been dissected and vindicated by scholars. Again and again, what seem to us fanciful stories from the *Histories,* such as gathering ledanon from the beards of billy goats or sheep whose tails are so large that they need wheeled carts to carry them, have been shown to be true: "The lesson is clear: condemning Herodotean tales out of hand because they do not seem likely may be more an indication of the parochial viewpoint of critics than of Herodotean naivety."[57] Regardless of these recent scholarly discoveries, we must admit that Herodotus does include in his *Histories* some fanciful stories that will never be vindicated, stories of giant maneating ants from India that guard deserts of gold or of black men whose semen is also black.

For his part, Lawrence includes both the story of an Arab youth who responds to incredible pain in a fantastically casual manner and his own exaggerated journey across the Sinai in forty-nine hours. Generally, however, Lawrence's narrative is free of tall tales, if not epic magnification. From the start, Lawrence announces his commitment to recording the events as they happened and admits the possibility of minor errors: "The record of events was not dulled in me and perhaps a few actual mistakes crept in—except in details of dates or numbers—but the outlines and significances of things had lost edge in the haze of new interests."[58] As he was writing *Seven Pillars of Wisdom,* Lawrence wrote to George Bernard Shaw of his commitments to historical accuracy: "To my astonishment, after peace came I found I was myself the sole person who knew what had happened in Arabia during the war: and the only literate person in the Arab Army. So it became a professional duty to record what happened."[59] And later, to E. M. Forster, Lawrence called his book a "complete narrative of what actually happened in the Arab Revolt."[60]

As a response to the mythic portrait of Lawrence of Arabia, critics such as Aldington and Mousa circled in with hopes of discrediting both Lawrence and his *Seven Pillars of Wisdom.* Since that critical frenzy, however, scholars such as Jeremy Wilson, Lawrence's authorized biographer, have done much to introduce a more balanced reading. Wilson has shown that while Lawrence was indeed prone to exaggeration and play—indeed Lawrence admitted as much—many of the criticisms of the veracity of his account suffer from poor knowledge of the appropriate sources.[61]

Applying Plutarch's criticisms of Herodotus to Lawrence has opened up a series of questions regarding whether and how we should understand Lawrence's literary enterprise as history, more specifically as something like ancient history, or more specifically still as something like Herodotus' own *Histories.* Many of Lawrence's critics assume that they understand exactly what sort of history Lawrence is writing and that they can thereby level at him criticisms appropriate to his endeavor. Much of this essay has been devoted to dismantling that assumption. The difference between Lawrence and Herodotus on the one hand and modern historians on the other, between an ancient history that is artistically composed and a modern history that aspires to be a science, is not that one records objective and verifiable facts and the other only fanciful stories. There is much in Herodotus and Lawrence that is true, objective, and verifiable. But that is not all that is in Herodotus and Lawrence. Mixed in with the objective and verifiable, governing the entire narrative, and permeating every description is a concern for a different truth. What is the character of this different truth? In a letter to Edward Garnett in 1922, Lawrence writes: "I'm glad you feel the veracity of the story. It was written in dead earnest . . . and I think it's all spiritually true." Both historical veracity and spiritual truth are woven together under the flag of history. We shall soon see these same themes developed as we turn our attention to Nietzsche, another modern philosopher of history. But whatever the exact character of this spiritual truth, when writing under its spell, when held by its rapture, both Herodotus and Lawrence are wont to say to their critics, bickering over the details: "Hippoclides does not care."

Nietzsche: On the Uses and Disadvantages of History for Life

In a letter to Lionel Curtis, dated December 1927, Lawrence writes: "One of the ominous signs of the time is that the public can no longer read history. The historian is retired into a shell to study the whole truth; which means that he learns to attach insensate importance to documents. The documents are liars. No man ever yet tried to write down the entire truth of any action in which he has been engaged. All narrative is parti pris. . . . *We know too much, and use too little knowledge*" (my emphasis).[62] This letter reads like a paraphrase of an early essay of Nietzsche's, entitled *On the Uses and Disadvantages of History for Life.* The extent and character of Nietzsche's influence on Lawrence has yet to be adequately treated in the scholarly literature. The best treatment to date is, not surprisingly, from the French, in Reneé and André Guillaume's *An Introduction and Notes: T. E. Lawrence's* Seven Pillars of Wisdom.[63] That this comparison has not been sufficiently made seems all the more strange when we consider that Lawrence compared his own *Seven Pillars of Wisdom* to, among other "titanic" works of modern literature, Nietzsche's *Thus Spoke Zarathustra.*[64]

Nietzsche begins *On the Uses and Disadvantages of History for Life,* the second of his four *Untimely Meditations,* with a diagnosis of his age, taking "certain striking symptoms" to suggest that "we are all suffering from a consuming fever of history."[65] History is figured as a disease because it has become for Nietzsche inimical to life. Life flourishes under the "unhistorical" sense, the capacity of an animal to forget, to live entirely in the moment free of past and future, with its horizons reduced to a near point. But man is the animal doomed to live in time, to remember and to anticipate, to struggle under the burden of the "historical" sense. Man is caught between his own Scylla and Charybdis, precariously perched between the historical and unhistorical senses. Too much of the unhistorical and man reverts to his blissful but animal origins; too much of the historical and man is paralyzed under the pressures of the limitless past and future, for "[f]orgetting is essential to action of any kind."[66] Man is the animal that cannot but live historically; but man is also the animal who cannot but choose *how* he will live historically.

There are, Nietzsche tells us, three basic species of history: the monumental, the antiquarian, and the critical. And these must be properly balanced: "[E]very man and every nation requires, in accordance with its goals, energies and needs, a certain kind of knowledge of the past, now in the form of monumental, now of antiquarian, now of critical history."[67] The problem with the modern age is not that it has become oversaturated with history, but that it has become consumed by the antiquarian mode of history, the history that belongs "to him who preserves and reveres."[68] The antiquarian historian sets himself the task of collecting all the facts and events of the past, however great or small, and arranging them in such an order that he can follow them forward to his own age. And while the antiquarian historian must almost always limit his own researches to a specific region or era—so as to collect and arrange *all* the facts and events—with the help of other antiquarian historians, he hopes one day to put in their place all the pieces of the great historical puzzle, to trace the "then" to the "now" in all its wondrous detail. By this exercise, the antiquarian historian situates himself in a long and distinguished "stream of becoming" and thereby explains and *justifies* his own existence.[69]

This is, according to Nietzsche, the motivation behind the practice of modern history: "it is this which is today usually designated as the real sense of history."[70] But the only purpose this antiquarian history serves, "the happiness in knowing that one is not wholly accidental and arbitrary but grown out of a past as its heir," is also a great danger. For insofar as the present is, for the antiquarian historian, simply the limit of history and the future an alien, even hostile venture, he "knows only how to *preserve* life, not how to engender it."[71] He preserves by tracing backward, by speaking of the roots of the tree of the present. And while the antiquarian historian does serve life by preserving it, he is also potentially hostile to life. Having thrown open the floodgates of the past, the antiquarian historian forces on modern man the impossible task of assimilating the "[h]istorical knowledge [that] streams in unceasingly from inexhaustible wells."[72] The result is a "uniquely characteristic 'subjectivity'": "the most characteristic quality of modern man: the remarkable antithesis between an interior which fails to correspond to any exterior and an exterior which fails to correspond to any interior—an antithesis unknown to the peoples of earlier times."[73] Modern man is unable to assimilate, unable to put to the service of life the swill of undifferentiated information shoveled into him by historians. As Lawrence writes to Curtis, "We know too much, and use too little knowledge." Modern man is a "walking encyclopedia," whose voluminous contents bears not at all on the binding and cover he shows the world. Too much history has made man a shadow of his former self.

Lawrence concurs with Nietzsche's diagnosis of the modern historical age, and testifies to his own characteristically modern subjectivity: "The two selves, you see, are mutually destructive. So I fall between them into the nihilism which cannot find, in being, even a false God in which to believe."[74]

And in his letter to Curtis, quoted above, Lawrence again echoes Nietzsche: "One of the ominous signs of the time is that the public can no longer read history." The public can read history only through an antiquarian lens, Lawrence suggests, for that is all that our historians are writing. It is Nietzsche's antiquarian historian whom Lawrence is lampooning when he goes on to write, "The historian is retired into a shell to study the whole truth; which means that he learns to attach insensate importance to documents."

And it is the antiquarian historian who harbors aspirations to make of his ancient art a science. For Nietzsche, this would amount to a betrayal of history, "for we want to serve history only to the extent that history serves life."[75] History cannot serve two masters. To which idol, then, shall history pray: to all that is noble and beautiful in life or to all that is objective and verifiable? "History become pure, sovereign science would be for mankind a sort of conclusion of life and a settling of accounts with it. . . . Insofar as it stands in the service of life, history stands in the service of an unhistorical power, and, thus subordinate, it can and should never become a pure science."[76] And yet Nietzsche suggests that a glance at this modern age reveals that his question has already been answered. Into the "constellation of life and history" has already penetrated "a gleaming and glorious star . . . *the demand that history should be a science.*" The result is clear: "Now the demands of life alone no longer reign and exercise constraint on knowledge of the past."[77]

That the scales of modern history have already decided against life suggests to Nietzsche that he and his meditations are "untimely"—a name with two important senses. First, that Nietzsche is "untimely" suggests that he resists his present age: "untimely—that is to say, acting counter to our time and thereby acting on our time and, let us hope, for the benefit of a time to come."[78] Contrary to so many who feel themselves to have been born too early or too late, for Nietzsche, to be untimely is not to while away one's days in the vicarious pleasures of studying the past or in dreamy speculation on the future, but to understand the present and to divine the future, and so to put the past to use in their transformation. To be untimely is to be living both in the past and the future, which brings us to the second sense. Not only does Nietzsche resist his age, but he testifies that he found the very resources for his resistance, the weapons with which he might lay siege to modern history, in his study of the past: "[I]t is only to the extent that I am a pupil of earlier times, especially the Hellenic, that though a child of the present time I was able to acquire such untimely experiences. That much, however, I must concede to myself on account of my profession as a classicist: for I do not know what meaning classical studies could have for our time if they were not untimely."[79] "[E]specially the Hellenic"—with these words Nietzsche invites us to compare his prescription for modern history with the ancient Greek historians. After all, the modern antithesis between interior and exterior was "unknown to the peoples of earlier times," so perhaps we might learn from them how better to do history. Nietzsche suggests that when we write untimely history, we look to the Greek historians, such as Herodotus, for guidance.

Modern history suffers from a dearth of monumental and critical approaches, Nietzsche suggests. While antiquarian history thrives on comprehensive detail, monumental history "pertains to [man] as a being who acts and strives"[80] and so highlights particular moments of the past: "That the great moments in the struggle of the human individual constitute a chain, that this chain unites mankind across the millennia like a range of human mountain peaks, that the summit of such a long-ago moment shall be for me still living, bright and great—that is the fundamental idea of the faith in humanity which finds expression in the demand for a *monumental* history."[81] The record of these rare and glorious achievements provides "the man of action" models of human excellence, reminds him "that the greatness that once existed was in any event once possible and may thus be possible again."[82] Paired with this positive presentation of the past is critical history, which pertains "to [man] as a being who suffers and seeks deliverance" and so devotes itself to the scrupulous examination and condemnation of the past. Why condemnation? First, because there is the paralyzing pressure of a limitless past, the need for a horizon wherein action is possible. Second, because there is much in humanity's past that is best forgotten, so much inimical to the flourishing of life in the present. Man suffers from the umbilical cord of his own history and seeks to be cut free from it.

When antiquarian, monumental, and critical histories are yoked together, there opens the possibility of what Nietzsche calls "genuine" history. That the present age is "oversaturated" with the antiquarian means that whoever wishes to resist the histories of the day will appear to his contemporaries as too monumental, too critical, and insufficiently concerned with the objective and the verifiable. This bravado is what Nietzsche calls for and what Lawrence takes himself to be answering. Not every one can be a great, genuine historian; only the "rarest minds" are worthy of this art. We should not heap derision on the "workmen who sift and carry merely because they can certainly never become great historians." But neither should we confuse them with great historians, nor hold great historians to those standards set by the "apprentices and handymen."[83] The genuine historian must be qualified for his subject matter, and not by publications and academic chairs: "He who has not *experienced* greater and more exalted things than others will not know how to interpret the great and exalted things of the past" (my emphasis).[84]

According to our common, antiquarian sensibility, experience does not contribute to objectivity, but rather pollutes it. If a historian is too involved in his research, this thinking goes, then he cannot possibly be objective. To be objective is to take up the perspective of a third and disinterested party, freed from all prejudice, desiring only truth and justice. But it is this easy association of objectivity with truth and justice that Nietzsche wishes to upset: "[T]he word means a condition in the historian which permits him to observe an event in all its motivations and consequences so purely that it has no effect at all on his own subjectivity: it is analogous to that aesthetic phenomenon of detachment from personal interest. . . . One goes so far, indeed, as to believe that he to whom a moment of the past *means nothing at all* is the proper man to describe it."[85] Rather than the standard by which great history should be judged, Nietzsche claims, objectivity is another face of modernity's fraught subjectivity. To aspire to a passionless and detached relationship to the past, one that does not make any claims on your present life, is to aspire to the separation of the interior self, crammed full of historical knowledge, from the exterior self, which is unable to transform and apply that historical knowledge to its life. So split, the objective subject fails to become a coherent self: "If the personality is emptied in the manner described and has become eternally subjectless or, as is usually put, objective, nothing can affect it any longer."[86] To aspire to objectivity is to aspire to a debased subjectivity.

Objectivity has therefore wrongfully claimed justice and truth as its own. Justice is not the annulling of all care for the subject at hand. That annulment is objectivity, but "Objectivity and justice have nothing to do with one another."[87] To be a just historian is to care deeply about the past you are recording, indeed to care so because the past offers a reflection of yourself: "only when you put forth your noblest qualities in all their strength will you divine what is worth knowing and preserving in the past. Like to like! Otherwise you will draw the past down to you."[88] The just historian dismisses objectivity, but desires truth: "truth, not as cold, ineffectual knowledge, but as a regulating and punishing judge; truth, not as the egoistic possession of the individual, but as the sacred right to overturn all the boundary-stones of egoistic possessions; in a word, truth as the judgment of humanity."[89] For the just and genuine historian, truth is indeed the standard, but not truth understood as objective and verifiable, "truth that eventuates in nothing."[90] The truth that motivates genuine history is the truth that serves life. Nietzsche makes this equation explicit quite early in the essay: "true, that is to say, life-enhancing."[91] By life-enhancing, Nietzsche does not mean the preservation of any life whatsoever, but the promotion of the "higher life," whatever "expand[s] the concept 'man' and make[s] it more beautiful," the transformation of "what they have learned into a more elevated practice."[92] There will be, however, even in just and genuine history, much that is true is the sense of objective and verifiable. But those objective and verifiable truths will be yoked to another purpose, "a desire to see new greatness emerge."[93] We suppose that this is what Lawrence had in mind when he wrote to Edward Garnett that his book was both historically veracious and spiritually true.

To this higher aim, Nietzsche claims, modern antiquarian history has shown a profound indifference: "It is precisely where the highest and rarest is to be represented that this ostentatious indifference becomes most frustrating."[94] In effect, modern antiquarian history, with its indifference to justice and truth, has said to life ου φροντις, that it does not care. And in response to *this* indifference to life, Nietzsche registers his own indifference, echoing young Hippoclides: "There are very many truths that are a matter of complete indifference; there are problems whose just solution does not demand even an effort, let alone a sacrifice. In this region of indifference and absence of danger a man may well succeed in becoming a cold demon of knowledge."[95] When Nietzsche professes his indifference, when Lawrence carves ου φροντις over his door, they are not renouncing the care of history but rather pledging themselves to it anew. They are distinguishing themselves from those whose care for history is a care for objectivity. They are committing themselves to care for history, to care that history make claims on our present lives. To the cold demons of knowledge, the antiquarian historians retired into their shells, the critics registering every infraction of objectivity, Nietzsche and Lawrence, standing on the table with Herodotus and Hippoclides, shout in reply, "ου φροντις!"

Notes

1. See George Rawlinson, trans., *The History of Herodotus* (New York: Tudor, 1928), VI.129, 349–50. I use Rawlinson's translation not because it is the most accurate or elegant, but because it was the translation Lawrence knew, although he undoubtedly knew the original Greek as well. See Lawrence's letter to David Garnett in David Garnett, ed., *The Letters of T. E. Lawrence* (London: Cape, 1938) 680–82. Rawlinson has rendered ου φροντις as "What does Hippoclides care?" I prefer the translation Lawrence uses most often, "Hippoclides does not care."
2. T. E. Lawrence to A. E. "Jock" Chambers, dated August 8, 1924, quoted in John E. Mack, *A Prince of Our Disorder: The Life of T. E. Lawrence* (Boston: Little & Brown, 1976), 29. Lawrence placed the lintel over the door in 1924. Hippoclides was already on his mind, as evidenced by this letter to Jock. Only in the summer of 1925, however, did Lawrence finally carve ου φροντις into the stone.
3. On August 30, 1922, Lawrence enlisted in the ranks of the RAF as "J. H. Ross." In January 1923, he was discharged from the RAF following press exposure. In March he enlisted in the Tank Corps as "T. E. Shaw." In August 1925, following threats of suicide, he was transferred back to the RAF,

retaining the name "Shaw." See the Chronology in Jeremy Wilson, *Lawrence of Arabia* (Phoenix Mill: Sutton, 1998), viii.

4. Paul Marriott and Yvonne Argent, *The Last Days of T. E. Lawrence* (Portland: Alpha, 1996), 165.
5. Robert Graves and B. H. Liddell Hart, *T. E. Lawrence to His Biographers* (Garden City: Doubleday, 1963), Vol. I, 81.
6. David Garnett, *The Letters of T. E. Lawrence,* 746.
7. Henry G. Liddell and P. G. Glare, eds., *A Greek-English Lexicon* (Oxford: Oxford University Press, 1995), 1956–57.
8. For a judicious discussion of Lawrence's various statements about the identity of "S. A.," see Jeremy Wilson, *Lawrence of Arabia: The Authorized Biography of T. E. Lawrence* (New York: Atheneum, 1990), 672–74.
9. T. E. Lawrence, *Seven Pillars of Wisdom* (London: Penguin, 1962), 23.
10. Plutarch, Lionel Pearson and F. H. Sandbach, trans., *On the Malice of Herodotus* (Cambridge: Harvard, 1965), 87.
11. Ibid., 2.
12. David Garnett, *The Letters of T. E. Lawrence,* 681. Lawrence suggested that David Garnett ask his younger brother, A. W. Lawrence, to take on the project. Arnold agreed and edited a beautiful edition for the Nonesuch Press.
13. Quoted on back cover of the Penguin Edition of T. E. Lawrence, *Seven Pillars of Wisdom.*
14. T. J. Luce, *The Greek Historians* (New York: Routledge, 1997), 17; see Herodotus, *Histories* IV.30.
15. T. J. Luce, *The Greek Historians,* 17.
16. I do not mean to suggest that Lawrence incorporated actual text from this previous manuscript. Indeed, that would have been impossible, for Lawrence destroyed this manuscript in 1914. What I do mean to suggest is that the vast knowledge Lawrence acquired through his early travels in the Middle East finds expression in *Seven Pillars of Wisdom.*
17. T. E. Lawrence, *Revolt in the Desert* (New York: Doran, 1927).
18. T. E. Lawrence, *Seven Pillars of Wisdom,* 9.
19. T. E. Lawrence, *Seven Pillars of Wisdom,* 23.
20. *The History of Herodotus,* 1.
21. T. E. Lawrence, *Seven Pillars of Wisdom,* 684.
22. See Friedrich Nietzsche, *On the Uses and Disadvantages of History for Life* (Cambridge: Cambridge, 1983), 69: "fame is something more than the tastiest morsel of our egoism, as Schopenhauer called it: it is the belief in the solidarity and continuity of all ages and a protest against the passing away of generations and the transitoriness of things."
23. *The History of Herodotus,* 1.
24. T. E. Lawrence, *Seven Pillars of Wisdom,* 9.
25. Ibid., 9.
26. Plutarch, *On the Malice of Herodotus,* 2–6.
27. Ibid., 21.
28. Ibid., 129.

29. Ibid., 23–25.
30. *The History of Herodotus,* 97.
31. See Herodotus, *Histories* II.119.
32. See Herodotus, *Histories* I.135, II.4.
33. Plutarch, *On the Malice of Herodotus,* 29.
34. Ibid., 129.
35. T. E. Lawrence, trans., *The Odyssey of Homer* (New York: Oxford, 1991), 162.
36. Richard Aldington, *Lawrence of Arabia: A Biographical Inquiry* (London: Collins, 1955), 12.
37. See Jeremy Wilson, *Lawrence of Arabia,* 433–34.
38. Suleiman Mousa, *T. E. Lawrence: An Arab View* (London: Oxford, 1966), vii.
39. T. E. Lawrence, *Seven Pillars of Wisdom,* 23.
40. Ibid., 24.
41. See Lawrence's letter to Vyvyan Richards, dated sometime in early 1923, in Malcolm Brown, ed., *T. E. Lawrence: The Selected Letters* (New York: Paragon, 1992), 225.
42. T. E. Lawrence, "Twenty-Seven Articles" published as Appendix IV in Jeremy Wilson, *Lawrence of Arabia,* 960–65.
43. Edward W. Said, *Orientalism* (New York: Vintage Books, 1979).
44. Malcolm Brown, ed., *T. E. Lawrence: The Selected Letters,* 224.
45. Dated June 10, 1927 (British Library). Quoted in Stephen Ely Tabachnick, *T. E. Lawrence Revised* (New York: Twayne, 1997), 74.
46. T. J. Luce, *The Greek Historians:* see especially chapters two and three. My comments on Greek history and historiography are much in debt to Luce.
47. Ibid., 4–5.
48. T. E. Lawrence, "Evolution of a Revolt," in A. W. Lawrence, ed., *Oriental Assembly* (London, 1939), 131.
49. The immediate antecedent of the pronoun "it" here is "the *Peloponnesian War.*" More generally, however, Lawrence is referring to the sort of written history that he, like Thucydides or Herodotus, composes.
50. T. E. Lawrence, *Seven Pillars of Wisdom,* 27; letter quoted in Stephen Ely Tabachnick, *T. E. Lawrence Revised,* 69.
51. A.W. Lawrence, ed., *T. E. Lawrence by His Friends* (London: Cape, 1954), 176.
52. See T. J. Luce, *The Greek Historians,* 21: "The noun and the verb (*historeo*) denote on-the-spot inquiry of what one sees and hears. In Herodotus it is the process or method of investigation rather than the result. Only in the fourth century and later did the word take on the specific meaning of 'a written account of man's past activities.'"
53. T. E. Lawrence, *Seven Pillars of Wisdom,* 25.
54. Ibid., 27.
55. Ibid., 22.
56. Richard Aldington, *Lawrence of Arabia,* 13.

57. T. J. Luce, *The Greek Historians,* 27–28. See Herodotus III.111–13.
58. T. E. Lawrence, *Seven Pillars of Wisdom,* 21.
59. David Garnett, *The Letters of T. E. Lawrence,* 356.
60. Ibid., 455.
61. Jeremy Wilson, *Lawrence of Arabia.*
62. David Garnett, *The Letters of T. E. Lawrence,* 559.
63. Reneé and André Guillaume, Hilary Mandleberg, trans., *An Introduction and Notes: T. E. Lawrence's* Seven Pillars of Wisdom (Oxshott: Tabard, 1998).
64. David Garnett, *The Letters of T. E. Lawrence,* 360, 467.
65. Friedrich Nietzsche, *On the Uses and Disadvantages of History for Life,* 59–60.
66. Ibid., 62.
67. Ibid., 77.
68. Ibid., 67.
69. Ibid., 62.
70. Ibid., 74.
71. Ibid., 74, 75.
72. Ibid., 78.
73. Ibid., 78.
74. Robert Graves and B. H. Liddell Hart, eds., *T. E. Lawrence to His Biographers,* Vol. I, 63.
75. Friedrich Nietzsche, *On the Uses and Disadvantage of History for Life,* 59.
76. Ibid., 67.
77. Ibid., 77.
78. Ibid., 60.
79. Ibid., 60.
80. Ibid., 67.
81. Ibid., 68.
82. Ibid., 69.
83. Ibid., 94.
84. Ibid., 91.
85. Ibid., 93.
86. Ibid., 87.
87. Ibid., 91.
88. Ibid., 88.
89. Ibid., 88.
90. Ibid., 89.
91. Ibid., 71.
92. Ibid., 75, 68, 71.
93. Ibid., 72.
94. Ibid., 93.
95. Ibid., 89.

Chapter Three

T. E. Lawrence and Historical Representation

James Gelvin

Contemporary American academic historians, such as myself, face a predicament that is partly of our own making. On the one hand, we bemoan the historical illiteracy of our students and the fact that what we do just does not seem to hold a great deal of interest for—or appear particularly relevant to the lives of—the public at large. On the other hand, all too many of us treat the very historical personalities and events that fascinate that public—the assassination of John F. Kennedy, the sinking of the *Titanic,* the identity of Jack the Ripper, the guilt or innocence of Lizzie Borden, and so on—with unconcealed scorn. At best, interest in these personalities and events is tolerated among our fellow professionals as guilty pleasures of a personal nature, seductive sideshows for slumming historians who might just as readily harbor a secret passion for Danielle Steel novels or television sitcoms. Or these personalities and events might be trotted out to perform the role of literary device, to be indulged because of their potential to whet a prospective audience's appetite for "real history." More often than not, however, when it comes to writing the dissertations and monographs that are the meat and potatoes of the contemporary historical profession, these personalities and events are, at present, only cursorily referenced—if, indeed, they are referenced at all.

T. E. Lawrence is one of those personalities whose public appeal and cult-like following, while providing grist for the popular history and psychobiography mill, baffles, troubles, and to a certain extent angers academic historians. But for that anger, it would be impossible to explain the delicious malevolence with which the late historian, essayist, and professional gadfly,

Elie Kedourie, incises Lawrence—and by implication the Lawrence cult—in the opening paragraph of his essay "Colonel Lawrence":

> On the threshold of the contemporary Middle East stands the figure of T. E. Lawrence, an object at once of awe and pity. He is a portent, a symbol of the power of Chance over human affairs, and of the constant irruption into history of the uncontrollable force of a demonic will exerting itself to the limit of endurance. The consequences of his actions have touched numberless lives, and yet their motives were strictly personal, to be sought only in his intimate restlessness and private torment. The poverty of his ideas matches only the passion with which he pursued their realisation. And the cruel irony of his fate is that, though he was intent chiefly on the salvation of the soul of T. E. Lawrence, a private person, it is as Colonel Lawrence, a public legend, engaged in a dubious adventure, that he may claim to survive in men's memories. And he needs must, further to embitter the irony, following his own principles and convictions, acquiesce in such a fate. For he was both a liberal and a romantic. As a liberal, he was blithely unconscious of the gulf that must yawn between thought and action, of the fatal change which comes over thought when flesh and blood enact its consequences; and as a romantic, he believed that the Kingdom of God can be taken by storm, and that political action is a passport to eternal salvation.[1]

Although Kedourie originally published this essay almost half a century ago, many academic historians of a later generation would readily concur with his skeptical assessment of Lawrence's intellectual attributes and personal qualities, and would perhaps throw in a complaint or two about Lawrence's cavalier approach to conventional historical methods and standards (as Kedourie also does later in his essay) for good measure. Indeed, even among historians inured to the present-day excesses of political correctness and the treacherous world of academic politics, Lawrence's overt racism (Semites "were a limited, narrowminded people, whose inert intellects lay fallow in incurious resignation"),[2] narcissism ("In these pages the history is not of the Arab movement, but of me in it"),[3] delusions of grandeur (it is easy to forget, reading Lawrence, that the Arab Revolt was a sideshow to a sideshow, representing little more than a flank of an army that was fighting in a secondary theater of a conflict to be won or lost in France), unabashed self-promotion (his fever-induced strategic epiphany—that the entente powers could bleed the Ottoman Empire by blowing up rail lines—had, in fact, been proposed earlier by a French naval intelligence officer and was circulating in military circles for months before Lawrence landed at the port of Jidda),[4] and imperialist fantastication (his musings about losing himself among faceless indigenes or passing for a near God among the childlike bedouin), are hard to swallow. Add to that the internal contradictions and inconsistencies that mar his works, his open abnegation of the strictures of responsible scholarship ("In this book also, for the last time, I mean to be my own judge of what to say"[5]), and his recourse to a prose style that seems, at times, to run closer to the prose style of Bulwer-Lytton than to that of Edward Gibbon (much less to the more pedestrian but overtly disinterested prose style employed by most academics writing today), and it is easy to see why historians might judge Lawrence's work both irritating and solipsistic—and certainly not a source to be taken without a mountain of salt.

Nonetheless, the very fact that Kedourie would expend his energies on writing an essay whose subject is T. E. Lawrence—even one that purports to unmask a secular saint—marks him as a product of an earlier era. For better or worse (and I believe the former), most contemporary academic historians show little enthusiasm for debates about whether or not Lawrence was a hero or fraud or whether he was an agent of history or pretender to that title. As far as they are concerned, Lawrence's run as leading man in the historical drama has ended, and they have already recast him as just another bit player or, more accurately, a member of the historical chorus. It is, in sum, not Kedourie's revisionism that separates him from his epigones. Rather, it is his unapologetic (but not entirely untroubled) adherence to a "great men" approach to history—an adherence that, while wholly consistent with the role Kedourie assumed as a self-styled political reactionary and academic provocateur, has become largely unfashionable among mainstream academic historians.

How has this come to pass? Because the reproduction of academic disciplines takes place through an apprenticeship system, knowledge institutionalized in the academy represents one of the few closed systems in which historians are still able to maintain that the idealist and progressivist assumptions of nineteenth-century intellectual history hold true. Thus, while it has become commonplace for historians to protest that privileging the activities of so-called great men in determining the course of history in the world at large is a historically determined narrative artifice that renders 99 percent of humanity voiceless and passive and thereby disenfranchises individuals from their own history, we nevertheless readily concede to ourselves the right to stand, as the cliché runs, on the shoulders of giants. As a result, the exploits of those giants have become fodder for countless graduate seminars through which young scholars are initiated into our guild.

The resulting historical narrative of itself begins with the "Annales School" of historians (so named after a French historical journal, originally called *Annales d'histoire économique et sociale,* to which many of the greatest French historians of the post - World War II period contributed), whose most famous practitioner, Fernand Braudel, taught his fellow historians that the "event"—the most obvious level of historical knowledge that has been the focus of so much historical attention (*l'histoire événementielle*)—is but a smoke screen blinding us to the dominant loci of the historical saga: the *structure,* that is, the environmental substratum whose transfiguration in the long term (*longue durée*) provides the context for all human activity, and the stratum of *conjoncture* wherein people organize their social lives and the reproduction of their societies.

While Braudel applied his three-tiered model to social history, others sought to apply its multi-layered and long-term, epochal perspective to understanding cultural change. A variety of scholars contributed, sometimes consciously, sometimes unwittingly, to this effort: French scholars of the "*mentalités*" school who, taking their cue directly from the original *annalistes,* applied their own vision of the *longue durée* to cultural stasis and transformation; British Marxists like E. P. Thompson who, by repudiating a priori categories ("base," "superstructure," "class") and by examining popular attitudes toward social hierarchy, rights, community, and law, breached the wall dividing social from cultural history; anthropologists such as Clifford Geertz and Claude Lévi-Strauss, who demystified culture and pioneered techniques for its interpretation; Marshall Sahlins, who explored ways to resolve the "dialectic" of anthropological fixity and historical change; and, of course, Michel Foucault, whose notion of discursive genealogies—and the stratum-by-stratum recovery of past orderings of knowledge and the structures of social power with which they were fused—posed a direct challenge to the gentlemanly game of intellectual history.

That this list is incomplete and perhaps a bit arbitrary (there are many others whose names might just as well have been included) is a testament to the vitality of the historical discipline over the past half century. But however incomplete and idiosyncratic the list, one can discern in it a definite trend that bears directly on the topic of Lawrence and Lawrence scholarship. With the emergence of the "new historicism" and the "new cultural history," historians have become more interested in the cultural milieux that produced so-called great men than in the great men themselves who, as it had heretofore been assumed, seemingly produced those cultural milieux and actualized them through innovative intellectual effort or revolutionizing political activity. Of particular interest to these new historicists and new cultural historians have been the descent, diffusion, and nature of commonly held assumptions that have guided the thinking not only of those great men, but of everyone else who lived at the same time and in the same place and who, more often than not, bore mute witness to the pervasiveness and determining effect of those inveterate assumptions through their praxis. In other words, rather than emphasizing that which made the so-called great men unique, the above-cited historians commonly look at that which rendered them products of their time, embodiments of that which Auguste Comte calls the "spirit of the age," or, more prosaically, their Zeitgeist.

Take Lawrence, for example. With all due respect for the adoration of Lawrence's contemporary acolytes, can it truly be stated that Lawrence's ideas were in any way unique, or that the unexpressed assumptions that underlay those ideas were any different from those held by his contemporaries such as Mark Sykes, Arnold Toynbee, William Henry Shakespear, Henry MacMahon, St. John Philby, Gertrude Bell, D. G. Hogarth (Lawrence's own mentor), or any number of British "Oriental experts," adventurers, diplomats, military officers, intelligence agents, and rogues whose opinions (not to mention derring-do) hardly register with today's public? Indeed, can it truly be stated that Lawrence's ideas would have appeared anything but commonplace to the typical Briton of privilege living during the apogee of British imperialist triumphalism and educated at public schools and Oxbridge?

There is, for example, Lawrence's embarrassing description of Semites, to which I have already alluded. Here is a more expanded rendering of the passage from which I extracted the quotation cited earlier:

> Semites had no half-tones in their register of vision. They were a people of primary colors, or rather of black and white, who saw the world always in contour. They were a dogmatic people, despising doubt, our modern crown of thorns. They did not understand our metaphysical difficulties, our introspective questionings. They knew only truth and untruth, belief and unbelief, without our hesitating retinue of finer shades.
>
> This people was black and white, not only in vision, but by inmost furnishing: black and white not merely in clarity, but in apposition. Their thoughts were at ease only in extremes. They inhabited superlatives by choice. . . . They were a limited, narrowminded people, whose inert intellects lay fallow in incurious resignation. Their imaginations were vivid, but not creative.[6]

Now, compare Lawrence's comments with the following statement on the same topic, this time taken from *Modern Egypt* by the Earl of Cromer, the first British consul general to Egypt, whose career predated that of Lawrence by little more than a generation. While differing in their choice of metaphors to describe the "Oriental other," both Lawrence and Cromer share the penchant to render the Oriental as a specimen of primitiveness to be placed under the microscope by the Western observer whose privileged position and (purported) sharply honed rationalism make him the font of scientific verities:

> Sir Alfred Lyall once said to me: "Accuracy is abhorrent to the Oriental mind. Every Anglo-Indian official should always remember that maxim." Want of

> accuracy, which easily degenerates into untruthfulness, is, in fact, the main characteristic of the Oriental mind.
>
> The European is a close reasoner; his statements of fact are devoid of ambiguity; he is a natural logician, albeit he may not have studied logic; he loves symmetry in all things; he is by nature skeptical and requires proof before he can accept the truth of any proposition; his trained intelligence works like a piece of mechanism. The mind of the Oriental, on the other hand, like his picturesque streets, is eminently wanting in symmetry. His reasoning is of the most slipshod description. Although the ancient Arabs acquired in a somewhat high degree the science of dialectics, the descendants are singularly deficient in the logical faculty. They are often incapable of drawing the most obvious conclusions from any simple premises of which they may admit the truth. Endeavour to elicit a plain statement of facts from an ordinary Egyptian. His explanation will generally be lengthy, and wanting in lucidity.[7]

Or take another example: Lawrence on the port city of Beirut and its inhabitants, the mongrel Levantines, individuals who in another age would have been lauded for their cosmopolitanism but to the imperial mind of the early twentieth century clearly represented cultural corruption, if not a eugenicist's worst nightmare:

> Beyrout was altogether new. It would have been bastard French in feeling as in language but for its Greek harbour and American college. Public opinion in it was that of the Christian merchants, fat men living by exchange; for Beyrout itself produced nothing. . . . Beyrout was the door of Syria, a chromatic Levantine screen through which cheap or shop-soiled foreign influences entered. It represented Syria as much as Soho the Home counties.[8]

And here is a quote from the periodical *Near East* on the same topic of Levantine degeneracy. Interestingly, it is cited in an essay written by the indefatigable Kedourie, not about Lawrence but about his contemporary, Sir Mark Sykes: "Throughout the centuries many people have come to the city [this time, Constantinople], the city of the Great Whore has sucked most of them in and spat them out Levantines—a people who are not a people, without patriotism, without honor, talking myriad tongues in jargon, the sole people of the world without virtue."[9]

This is not to say that the mental topography that linked Lawrence with the "Oriental experts," adventurers, diplomats, military officers, intelligence agents, and rogues that made up his cohort was undifferentiated. This, of course, was hardly the case: some of this cohort treated the "natives" they encountered with undisguised contempt, others romanticized them, and still others (like Lawrence) flitted from pole to pole. The "Oriental experts," adventurers, diplomats, military officers, intelligence agents, and rogues also disagreed among themselves over the details of the grandiose plans they hatched to "awaken the East": Sharif Husayn or Ibn Saud? Zionists or Arabs? Semites or Turks? Damascus or Cairo? Noble Bedouin or renegade military officers? Nevertheless, they took it upon themselves to hatch those grandiose plans, and underlying those plans was a commonality of presuppositions derived from training, profession, social class (real or ventured), presumptive authority, and the circumstances of empire. Had Lawrence and Philby sat down over brandy and cigars in a London gentleman's club, they surely would have argued—but just as surely they would have grounded their arguments in a common set of assumptions.

Two of these assumptions are particularly relevant to our understanding of the designs and activities of both the stewards of empire and their counterparts among the eccentric travelers who roamed the Middle East in search of adventure. The first assumption, common among educated Europeans since the Enlightenment, was that there are universally applicable rules that guide the course of history. Foremost among these rules is the idea that there exists something called "civilization," and that, in the words of Robert A. Nisbet, this "civilization has progressed in the past, is now progressing, and will continue to progress into the illimitable future."[10] However one may have felt about the blessings or deficiencies of this civilization, whether one was a self-professed "modernist" or his "romantic" complement, the very fact that universal rules exist meant not only that the same rules for social organization should govern Britons and Indians and Arabs and even Frenchmen, but that there are universal standards for measuring civilizational attainment—a civilizational yardstick, if you will, calibrated according to the idealized self-conception of the average educated European and the secular nation-state ideal that Europeans professed. Armed with their knowledge of these rules, the "Oriental experts," adventurers, diplomats, military officers, intelligence agents, and rogues of empire felt no compunction about dragging the benighted natives they encountered into History and thus guiding them to their—and the rest of mankind's—destiny. The post - World War I mandates system, which made the "tutelage" of "peoples not yet able to stand by themselves under the strenuous conditions of the modern world" the "sacred trust of civilisation,"[11] represents perhaps the purest public expression of this universalist ethic and the civilizing impulse.

While the first assumption therefore construes the peoples of the world as divided into two categories—those who are part of an advancing civilization and those who are in need of civilizing—the second relevant assumption refrains from such categorical judgments. In fact, the second assumption seemingly stands in direct contradiction to the first. "Mankind," in the words of the German philosopher of history (and Lawrence contemporary), Oswald Spengler,

> has no aim, no idea, no plan, any more than the family of butterflies or orchids. . . . I see, in place of that empty figment of *one* linear history which can only be kept up by shutting one's eyes to the overwhelming multitude of the facts, the drama of *a number* of mighty Cultures, each springing with primitive strength from the soil of a mother-region to which it remains firmly bound throughout is whole life-cycle; each stamping its material, in *its own* image; each having *its own* idea, *its own* passions, *its own* life, will and feeling, *its own* death.[12]

For Spengler and like-minded romantics, the world thus consisted of diverse, incommensurable, culturally defined nations. Each nation's defining culture is but an outward manifestation of (to borrow a phrase from French philosopher Henri Bergson, another Lawrence contemporary) its unique élan vital—its vital force—that gave rise to that nation, rouses that nation, and, in effect, provides that nation's reason for being. If, however, a culture opens itself up to alien influences, that élan vital is diluted and, if that dilution continues uninterrupted, the nation is destroyed.

The notions that there exist both universal standards of civilization and a plethora of distinct cultures might seem at the outset irreconcilable, but reconciled they can be and often are. That a reconciliation of these two notions lies at the root of nationalist ideologies, for example, has inspired scholars of nationalism to refer to those ideologies as "Janus-faced": on the one hand, nationalist movements embrace Enlightenment and post-Enlightenment rationalism and its progressive and universalizing pretensions; on the other hand, nationalist movements represent themselves as the heirs to an ancient and distinctive national history that not only justifies the nation's right to a sovereign existence but that establishes each nation's contribution to modernity as unique.

The "Oriental experts," adventurers, diplomats, military officers, intelligence agents, and rogues who wandered the Middle East on behalf of empire or in spite of it achieved a similar reconciliation. This should not be surprising, given that their aim was, after all, to effect a national awakening, albeit the awakening of a nation that was other than their own. In other words, having found in the exotic lands they travelled a natural simplicity and *Gemeinschaftlichkeit* (organic solidarity) that eluded them at home, the more adventurous among them sought to become the agents who would effect the retrieval or strengthening of their chosen nation's lost or weakened élan vital—its raison d'être—and in the process guide the object of their devotion to its rightful place within the community of nations. This was the task T. E. Lawrence, as a "liberal and romantic," to borrow a phrase from Kedourie, shouldered. About this task Lawrence was far from reticent: "I had dreamed, at the City School in Oxford," he wrote in his afterword to

Seven Pillars of Wisdom, "of hustling into form, while I lived, the new Asia which time was inexorably bringing upon us."[13]

For Lawrence (as for many others in his cohort), those who peopled the Asiatic Arab provinces of the Ottoman Empire possessed an essence which might be called their "Arabness." While this Arabness was inherent to the inhabitants of the region, it was dormant and unself-conscious and thus needed to be awakened so that those inhabitants, too, might join the march of civilization. But of all the social strata inhabiting the Middle East, it was the Bedouin of the Arabian peninsula who were to play the principal role in bringing about the Arab revival, because the Bedouin were less affected by the corruption brought on by the intrusion of alien cultures and because they were on more intimate terms with the élan vital of the Arab nation. This, for Lawrence, was demonstrated by their very primitiveness, their appreciation for the Arabic language, and the simplicity of their religious beliefs. Indeed, while Lawrence believed that all Semites were, by their nature, the "monopolists of revealed religion," it was the Bedouin alone, the purest of their race, who roamed the desert where they "came near God."[14]

On the other hand, Ottoman Turks, mongrel Levantines, and effete town-dwellers in general represented for Lawrence sources of corruption. Once a great race, overlords of an empire that stretched from Persia to Spain, the Arabs had been defiled and the singularity of their vision splintered. With the strength of its unifying essence sapped, the Arab nation cleaved along a series of fractures: desert against sown, Muslim against Christian, Sunni against Shi'i, north against south. What the Arab race needed was a leader who could bring together the fragmented inhabitants of the region under the umbrella of their inherent Arabness. With Amir Faysal as his native interlocutor, and Faysal's bedouin as his foot soldiers, Lawrence would be such a leader.[15]

Lawrence took this foundation of tacit and express assumptions and constructed upon it a literary superstructure whose bricks and mortar consisted of carefully vetted vignettes and tendentious character sketches. He shaped those vignettes and character sketches into a narrative arc whose rise traced his own military and personal triumphs and whose declination followed a course of betrayal and perfidy to a point of final, tragic denouement. But whatever the internal cohesiveness of the narrative arc, the ultimate test of its strength would reside in the resilience of its foundation. Pressing into service the deconstructionist techniques championed by postmodern and postcolonial theorists, who have been more than willing to poke at the joists connecting narrative form and historical judgment, assumption and fact, and power and knowledge, academic historians have tested that foundation and determined it to be rickety indeed.

Thus, as might be expected, examples of Lawrence's expectation-induced delusions and misinterpretations abound in his writings. Let us start with his views of the Ottoman state and society. By 1916, the Ottoman Empire had ruled over the Arab Middle East for four hundred years. According to Lawrence, throughout their history Semites have exhibited the unfortunate tendency to oscillate between a state of delirium and a state of lethargy. The Turks, having had the good fortune to catch the Arabs in the latter state, were able to insinuate themselves into the Middle East with disastrous result: "Early in the Middle Ages, the Turks found a footing in the Arab states, first as servants, then as helpers, and then as a parasitic growth which choked the life out of the old body politic. The last phase was of enmity, when the Hulagus or Timurs sated their blood lust, burning and destroying everything which irked them with a pretension of superiority. . . . By stages the Semites of Asia passed under their yoke, and found it a slow death."[16]

Like all other empires, the Ottoman Empire did, in fact, support its vast imperial bureaucracy by extracting surplus from its subjects in the form of taxes and tribute. It is therefore indeed possible to make the case that, with regard to those who produced that surplus, the Ottoman imperial apparatus was a "parasitic growth," just as the ruling apparatuses of all other empires have been. But clearly Lawrence is neither preaching an anti-statist or anti-imperial message, nor is he concerned here with matters of political economy; rather, his aim is to make the case that the "Turks" (and Lawrence's confusion between the ethnolinguistic category "Turk" and the political category "Ottoman" is particularly telling) represented an alien presence that sapped the lifeblood from the "Arab nation." Inasmuch as Lawrence was projecting backwards notions of "Turks" and "Arabs" that did not exist in the region until the end of the nineteenth century (and even then only among a thin stratum of the population), such a view can only be regarded as absurd.

The inhabitants of the Ottoman Empire of course understood that some of their fellow subjects spoke Turkish as a native language while others spoke Arabic or Kurdish or Greek or whatever. They also understood, if they cared to delve into such matters, that the multifarious customs practiced in the Bosnian or Albanian or Anatolian provinces of the empire often differed from those practiced in the Arab provinces (although the significance of these differences was to a great extent mitigated by the fact that all inhabitants of the empire were subject to the same imperial strictures). There are even examples of medieval Arab literati attempting to pigeonhole the Turkish warriors entering the Middle East from the north, ascribing to them, for example, an inborn aptitude for archery and horsemanship. But language and customs are one thing, and political identity is quite another. The same goes for the recognition of difference and the assertion of an explicitly modern conception of primordial and immutable ethnic identities demanding national inscription. In point of fact, the very notion of the existence of a singular "Arab" identity correspondent to Lawrence's usage is belied by the very meaning of the word *ʿarab* among the native Arabic speakers who used it: until the very end of the nineteenth century, the word not only did not refer to a member of the dominant ethnolinguistic community inhabiting the Asiatic and North African provinces of the empire, it was commonly used as a term of contempt by town-dwellers when referring to the very "savage" Bedouin whom Lawrence championed.

But Lawrence was wrong in another way when he compares the Turks to a parasitic growth, for not only is he dividing Turk from Arab in a manner that is anachronistic, he is dividing the Ottoman structures of governance from the population in a manner that is wholly misguided. By the early twentieth century, there was hardly an aspect of economics, social structure, or politics in the Arab Middle East that had not been touched or even shaped by four hundred years of Ottoman rule. In effect, the nineteenth-century Ottoman state had created the economic playing field upon which all imperial subjects competed. The Ottoman state expanded security in rural areas, promulgated uniform commercial codes, devised strategies for industrial and agricultural development, broke down regional trade barriers, recognized private ownership rights to land, and constructed infrastructure (including the very railroads and telegraph lines Lawrence was to blow up) that bound together the far reaches of empire. From Damascus to Baghdad, late nineteenth-century Arab urban elites, whose very rise had been enabled or underwritten by Ottoman imperial policies, sent their children to Ottoman schools, spoke Turkish as well as Arabic, and marked their cosmopolitanism by wearing Ottoman fashions (such as the fez). Politically, those same elites vied among themselves for imperial recognition and patronage and, almost to a man, understood that the best hopes for their future lay within the framework of a rejuvenated Ottoman state—a project to which many of those elites applied themselves with alacrity.

If those same elites subscribed to any nationalism at all, it was overwhelmingly to Ottoman nationalism, or *Osmanlilik,* which had been promoted by the Ottoman government since the late 1830s and which, by the beginning of the twentieth century, they held in any one of a number of its permutations. As far as the notion of an Arab ethnic or linguistic nationalism was concerned, a vast majority of early twentieth-century Arab urban elites who considered such matters rejected it. As a matter of fact, most rejected the idea of Arab nationalism on the grounds that it would rend an empire already under siege and expose it to the intrigues of European imperialists—i.e., people like Lawrence. This is in fact just what happened.

It is true that by the beginning of the twentieth century there was the occasional political organization that advocated Arab autonomy and later, independence. But these organizations were few and far between and, ironically, located in the very cities and among the very Levantines and effete town-dwellers Lawrence and his "Arab" followers treated with contempt. Lawrence makes a big deal out of one of those organizations—al-Fatat (or the Fetah). In the *Seven Pillars* he writes: "Greater than the Ahad [the first such organization Lawrence describes] was the Fetah. . . . With the noisy facility of the Syrian—an apelike people having much of the Japanese quickness, but shallow—they speedily built up a formidable organization."[17] From reading Lawrence, it would be impossible to know that the total membership of al-Fatat from the time of its founding in 1909 or 1911 (depending on the source) to 1918 was about seventy. But Lawrence had his own reasons for exaggerating the importance of al-Fatat: when Lawrence's protégé, Amir Faysal, attempted to establish his political control over Syria in the aftermath of World War I, he turned to al-Fatat, which he himself had joined during the war, for help. The result was disastrous, both for al-Fatat and for Faysal's government. For its part, al-Fatat devolved into an organization whose raison d'être became dispensing political patronage. Since it admitted those who did not care about its nationalist message per se, but joined only to gain access to Arab government largesse, the stature the organization had acquired as the disinterested advocate for Arab nationalism declined precipitously. For his part, Faysal's reliance on the small clique that dominated the organization to rule insulated him from the increasingly restive population of Syria.[18]

Whatever the failings of al-Fatat during the postwar period, the organization did function as an outlet for the nationalist impulses of its members before the war. It would be a stretch, however, to attribute the same nationalist impulses to Faysal's father, Sharif Husayn. If there was an area of the Arab world in which the modern conception of nationality was especially feeble, it was the Arabian peninsula. Contrary to Lawrence's assertions, the British did not use the Arab nationalist card to enlist Sharif Husayn's support for the Arab Revolt. They did not need to: by the time the British made their approach to the sharif, he was more than willing to turn against the imperial government.

Under the Young Turks who had seized power in 1908, the Ottoman government exerted every effort to strengthen its authority so that it might resist renewed internal and external challenges, such as those concurrently at play in Libya, Yemen, and the Balkans. Strengthening the central government meant, of course, reducing the power of local potentates and power brokers, such as Sharif Husayn. By the outbreak of war, the sharif had already clashed with the Ottoman central government over a number of issues. Particularly galling to the sharif was the government's plan to extend the Damascus-Medina rail connection to Mecca, which, if effected, would not only have greatly augmented Ottoman control over the Hijaz, it would have ended the lucrative protection racket through which the sharif and his Bedouin clients extorted money from pilgrims to the holy city. Only after the Ottoman government began plotting his removal did the sharif purportedly became a nationalist, but even then it was a strange type of nationalism: he threw himself into the arms of an imperialist power, mobilized his forces through ties of clientage, and deployed his son to lead a rebellion that he had launched in the name of religion—not nationality. The national commitment of his retainers and clients also seems to have been of a strange type, for the size of the army fielded by his son varied in direct proportion to the quantity of British gold in its coffers.[19]

Nevertheless, Lawrence needed the Arab Revolt to be the vanguard of a national awakening. Thus, Lawrence made it that vanguard. In the *Seven Pillars,* Lawrence establishes the credentials of the sharif as follows: "Hussein, as politician, as prince, as Moslem, as modernist, and as nationalist, was forced to listen to their appeal. . . . In old age ambiguity covered his every communication. Like a cloud it hid his decision of character, his worldly wisdom, his cheerful strength. . . . Of religious fanaticism there was little trace. The sherif refused in round terms to give a religious twist to his rebellion. His fighting creed was nationality."[20] While none of Lawrence's assertions about Sharif Husayn's personality appears to have been true, it would hardly have suited Lawrence's purposes to present the sharif as many others who had to deal with him saw him: a narrow-minded old crank with a propensity toward paranoia; an opportunist who showed little enthusiasm for nationalism or other modern conceits but who was more than willing to aggrandize himself at the expense of potential rivals. Here is the sharif in the words of the Islamic modernist Muhibb al-Din al-Khatib, who edited the official sharifian newspaper, *al-Qibla,* at the beginning of the war but who soon abandoned Mecca and the sharif for the greener pastures of wartime Egypt: "[We] wanted all the best for King Husayn and his sons, on the condition that they would want all the best for the nation which they would lead. But King Husayn and his sons lived and died with the beliefs of the age of feudalism. They believed that the nation is the domain of the king."[21]

By the time the Arab army reached Damascus, Ottoman officials and their German advisors had already withdrawn, and the notables of the city had already organized an indigenous government that included representatives of some of the most respected local families. In his *Seven Pillars,* Lawrence does allude to this interlude, but is so blinded by his preconceptions and the constraints imposed by his own mythogenic fancies that he is unable or unwilling to acknowledge the role played by those preconceptions and constraints in setting off the tragic chain of events that undermined his entire project. The sharifians, under Lawrence's guidance and with British government acquiescence, destroyed what had been the government of Damascus and established one of their own. The British kidnapped Saʿid al-Jazaʾiri, whom the notables of the city had appointed "President of the Syrian Arab government," exiled him to

Haifa, and, more than likely, ordered his brother, ʿAbd al-Qadir al-Jazaʾiri, who had been appointed commander of the Arab cavalry before their arrival, killed.[22] Lawrence ensured that Faysal would be installed in their place. In the process, Faysal engendered the enmity of many prominent Damascenes who deeply resented him, his Hijazi ways, his uncouth Bedouin entourage, and his British friends. By 1919 the ranks of the anti-Fayal, anti-imperialist opposition had been swelled by non-elite Syrians who were exasperated by Faysal's seeming complicity in imperialist machinations, his inability to deliver on the popular demand for a Syria independent and undivided, and his titular leadership of a government marked by corruption and inefficiency—a government that had been imposed upon them by force of arms. Crowned king of Syria in March 1920, Faysal reigned but never ruled over an increasingly fractious political scene until, a little less than five months after his coronation, a French army brought closure to what he and his entourage gamely referred to as the fourth year of the Arab Revolt.

In the end, the Arab Revolt hustled no "new Asia" into form, nor did it have much of any lasting impact, save to unite Lawrence aficionados in the West and nationalist historians in the Arab Middle East around a common (but relentlessly naive) tale that begins with heroic struggle and ends with promises broken and dreams betrayed. This, of course, was to be expected: as described above, Lawrence's narrative attended so closely to the precepts of nationalist historiography and provided such a powerful rationalization for the deficiencies and shattered dreams of the nationalist elites of later generations that they could not but find it tailor-made to fit their own mythopoetic purposes. Like so many others in his cohort, Lawrence had blundered into an unfamiliar world guided only by a roadmap that was wholly misleading but of whose accuracy he was cocksure. And like so many others who used that roadmap when tracing the paths of their own narratives—from the all-but-forgotten Stephen P. Duggan, John de Vere Loder, Richard Coke, and Elizabeth P. MacCallum, to the still influential George Antonius and his countless Middle Eastern disciples—Lawrence neither redrew that roadmap nor deviated from it in the least.

But in the end, is it fair to expect him to have done so? Lest the contemporary reader be tempted to approach the shortcomings of Lawrence and those others with *Schadenfreude* or smugness, I would urge that reader to think of the contours marked on our own roadmaps, whether they might demarcate such post - Cold War pieties as "globalization" and "the clash of civilizations" or be laid out within a framework of "democratization theory." Perhaps, then, in the end the most appropriate way for the contemporary reader to approach Lawrence and the entire Lawrence phenomenon is as a cautionary tale, one that conjoins the impulse toward historical certainty with the perils of finding it.

Notes

1. Elie Kedourie, "Colonel Lawrence," in *England and the Middle East: The Destruction of the Ottoman Empire, 1914 - 1921* (Boulder: Westview Press, 1987), 88.
2. T. E. Lawrence, *Seven Pillars of Wisdom: A Triumph* (New York: Anchor Books, 1991), 38.
3. Ibid., 24.
4. See Eugene L. Rogan, *Frontiers of the State in the Late Ottoman Empire: Transjordan, 1850 - 1921* (Cambridge: Cambridge University Press, 1999), 218–19.
5. Lawrence, *Seven Pillars,* 26.
6. Ibid., 38.
7. The Earl of Cromer, *Modern Egypt* (New York: The Macmillan Company, 1908), 2: 146–47
8. Lawrence, *Seven Pillars,* 333.
9. Kedourie, "Sir Mark Sykes," in England and the Middle East: The Destruction of the Ottoman Empire, 1914 - 1921 (Boulder: Westview Press, 1987), 75.
10. Robert A. Nisbet, *Social Change and History: Aspects of the Western Theory of Development* (New York: Oxford University Press, 1969), 104.
11. J. C. Hurewitz, *The Middle East and North Africa in World Politics: A Documentary Record,* 2 vols. (New Haven: Yale University Press, 1975), 2:179.
12. Oswald Spengler, *The Decline of the West,* trans. Charles Francis Atkinson (New York: Alfred A. Knopf, 1945), 1: 21.
13. Lawrence, *Seven Pillars,* 661.
14. Ibid., 39, 40.
15. The "Faysal" mentioned here is the same as the "Feisal" mentioned in earlier chapters. Gelvin employs a different transliteration scheme than did Lawrence, hence the different English renderings. Indeed, Lawrence himself employed no single scheme, and so often the same name is spelled differently throughout *Seven Pillars of Wisdom.* Those concerned about these inconsistencies might recall Lawrence's infamous comment on transliteration in the Preface to *Seven Pillars of Wisdom:* "Arabic names won't go into English, exactly, for their consonants are not the same as ours, and their vowels, like ours, vary from district to district. There are some 'scientific systems' of transliteration, helpful to people who know enough Arabic not to need helping, but a washout for the world. I spell my names anyhow, to show what rot the systems are." *Seven Pillars of Wisdom,* 19 (Editor's Note).
16. Ibid., 44. Hulagu, by the way, was not a "Turk" but a Mongol. Timur was only part "Turk," on his paternal side; his maternal ancestry was Mongol.
17. Ibid., 47.
18. See James L. Gelvin, *Divided Loyalties: Nationalism and Mass Politics in Syria at the Close of Empire* (Berkeley: University of California Press, 1998), particularly 55–64.

19. See C. Ernest Dawn, *From Ottomanism to Arabism: Essays on the Origins of Arab Nationalism* (Urbana: University of Illinois Press, 1973), 1–53, 69–86; William Ochsenwald, "Ironic Origins: Arab Nationalism in the Hijaz, 1882–1914," in *The Origins of Arab Nationalism,* ed. Rashid Khalidi et al. (New York: Columbia University Press, 1991), 189–203; Hasan Kayali, *Arabs and Young Turks: Ottomanism, Arabism, and Islamism in the Ottoman Empire, 1908–1918* (Berkeley: University of California Press, 1997), 144–73.
20. Lawrence, *Seven Pillars,* 50.
21. Muhibb al-Din al-Khatib, autobiographical manuscript kept in the Salafiyya Library in Cairo and in Markaz al-watha'iq al-tarikhiyya in Damascus, 43.
22. Lawrence calls ʿAbd al-Qadir al-Jaza'iri "an Islamic fanatic, half-insane with religious enthusiasm and a most violent belief in himself" (*Seven Pillars,* 390). As with his assessment of Sharif Husayn, Lawrence is presenting an opinion that is not borne out by the facts. A descendent of a family that is still honored by Damascenes for their role in protecting Christians during the 1860 intercommunal riots (as Lawrence himself recounts), al-Jaza'iri was no doubt hated by Lawrence because he represented a threat to Lawrence's role as kingmaker.

Chapter Four

Reflective Heroes

Self-Integration through Poetry in T. E. Lawrence and Homer's Odysseus

Maren Cohn

Introduction

Our tale begins with a man who fights a grueling tribal war in a foreign country, a war in which he excels as much by ingenuity as by strength of arms. With the fighting over he begins to question both the military system that guided his efforts and his own moral involvement in the war, and he embarks upon on a long quest for self and home. After spending many years in lonely reflection he arrives anonymously in a new place. Here the man jealously guards the secret of his identity, but soon his name is dramatically revealed. He now also tells the story of his past adventures, a story that has no available witnesses: the entire account depends on his word. As it turns out, the man possesses no great attachment to the ideal of literal truth. No one matches him in using words to win a desired effect. Sometimes he contradicts himself, and even when he doesn't he frequently tells different versions of his exploits to different people. Besides his expertise as a storyteller the man is adept at disguise and gives himself a series of fictitious names. Such machinations of fiction and disguise strike many observers as questionable, even despicable, yet they serve two important aims. First: the man has become extremely famous; he hears tales about himself everywhere—some true, but some false. People tell stories about him for fun and for profit. His own stories both counteract and advance this process. They advertise his brilliant deeds while cloaking his person in mysterious ambiguity. He becomes as famous for misdirection and disguise as for boasting self-revelation, with the result that no one can be certain of the truth about him. Second: the tales operate internally, helping him to synthesize and comprehend the depths of his own experience through the poetic transformation of self. The first aim serves the second: only by clearing the slate of foreign accretions is the man free to explore and recreate the self. Of whom do we speak? Of two men: Homer's Odysseus and T. E. Lawrence. With the poetic activity of self-creation, however, their stories diverge and must be treated separately.

Odysseus is the hero of a highly complex, internally probing, self-conscious poem. In the *Odyssey*'s many scenes of storytelling and song its poet examines what it means both to tell tales and to hear them. The suitors listen unreflectingly to Phemius's song of the Achaeans' disastrous return, while the same words draw Penelope from her room in tears, begging the singer to stop (1.325; 44). Nestor says he could not tell the story of the war even in five or six years, for they fought in Troy for nine—as if to tell the tale would mean to relive it in real time (3.115; 19). Helen serves up *nepenthe* before urging her company to tell stories: "Whoso swallowed it mixed within his cup will not on that day let one tear down his cheeks, not though his mother and father died, not though men hacked to death his brother or loved son with the cutting edge before him and he seeing it with his eyes."[1] The apparently gracious Phaeacians think that wars are fought for the sake of the entertainments they produce (8.579; 80). The list would only become tedious, like Nestor's nine-year tale, if allowed to accumulate further. In the midst of the poem's meditation on the poetic process, the narrative Odysseus gives in books 9 through 12 assumes central importance. By its means, this man, who has lived violently and suffered profoundly for twenty years, grounds the remaking of himself into a husband, father, and king.

In laying before us its wide array of stories, tellers, and listeners, the *Odyssey* shows itself to be the product of introspection. But it does not engage in meditative reflection as we know it, nor does it portray its hero in the throes of contemplation. Homer's poem is equivocal and uncommitted; it forces its reader to shift constantly between conflicting points of view, and it offers a complex moral vision of a complex world. But the prior thought and reflection that crafted it did not lead it to become divided against itself. Its story concerns ambiguous characters; its ending leaves their futures unknown; it offers sometimes two or more versions of a single event, calling into doubt the human ability to know the world and its inhabitants, including the self. But it does not finally argue the absence of centers to be known. The poem stands solidly against the same disorienting vagaries of time and place that assault its hero and yet fail to destroy him completely. Odysseus can act in the end from a poetically known and integrated self, because there is such a self to underpin his action, because he has engaged in a synthesizing poetic process that has made such a self available to him.

Similar forces of uncertainty attacked Lawrence, whose thought-ridden habit eventually, inevitably inhibited integrated action. Although he embarked upon active war service with exalted ambitions for heroic action on epic and chivalric models, his actual experience finally presented only the grim choice between abandoning action altogether or acting falsely, obsessively, even pathologically, by dint of a will unsupported from within. Forced by circumstance to adopt the latter path during the war, Lawrence later, by enlisting in the Royal Air Force, chose the path of withdrawal into a sphere of mechanical activity inexpressive of an underlying individuality. His experience in the artistic realm followed a similar course, as he began work on *Seven Pillars of Wisdom* with ambitions that rivaled those with which he had entered the war. He intended the book to exhibit the same greatness of spirit that *The Brothers Karamazov, Thus Spake Zarathustra* and *Moby Dick* had shown before it,[2] and he struggled while writing it to find in the poetic process a means to attain and express the requisite sublimity of soul. But instead he discovered that he was not finally an artist, whether out of constitutional inability to lose himself in poetic inspiration or out of stubborn refusal to believe in art's redemptive powers. In any event, Lawrence could not allow his art to present him as, much less make him into, a unified hero of action in which he himself did not believe. He wrote not to weave the single fabric of an integrated self out of conflicting traits and impulses, but rather to communicate precisely those inner divisions. Echoing his decision in the field of action, he turned in poetry[3] to what he saw as the merely mechanical wordplay of translation and the literary equivalent of "objective" photography in *The Mint,* his record of life in the ranks.

Lawrence presented the pursuits of his later life as attempts to lose himself in the serviceable anonymity of the machine. Not all of the "mechanical" activities he undertook, however, turned out to fulfill their stated aim. Translation by its very nature implies not only selfless subordination to the original author, but also, paradoxically, the self-assertion that inheres in every new creation. When Lawrence encountered the great, challenging mirror of Homer's *Odyssey,* he produced a document that displays the residues not of a safe mechanical process but of convoluted dynamics between Lawrence and both Odysseus and Homer, dynamics that repeat the worst aspects of the translator's previous experience in war and art. *The Mint,* too, although for different reasons, similarly failed to achieve for Lawrence the grounding synthesis of self that Odysseus had attained both internally and communally.

Lawrence did, however, achieve a kind of peace in the RAF. If the aspiration after the heroic grandeur of a sublimely integrated soul was lost forever, in its place he managed to construct, in his life as boat designer and mechanic, something quietly beautiful. Here the comparison with Odysseus still holds, who was emphatically an earthbound hero and not one to give himself over to spiritual sublimity—if such an idea could even have occurred to him. But Odysseus' home life furnishes him a deeper and stronger support than Lawrence ever found in the RAF. The stability of the modern figure remained precarious: it was not the powerful centeredness one sees in Odysseus or, say, in a religious hero like Jesus, but rather the temporary quiet of an ocean whose depths continue to roil—if unobserved, yet vaguely felt by the consciousness of the surface. It thus becomes difficult to distinguish Lawrence's apparent peacefulness from despair. If he reached an integration of sorts, it was only an ironic one, a self-made raft launched out upon the abyss, a raft whose sailing may divert attention from the chaos below but does not found a solid structure above.

We have then before us an inquiry that extends in a number of directions. The first section of this essay treats Odysseus' experience while exploring the view that heroism arises from the integration of self and that poetry is the means of achieving that integration. The second section looks at Lawrence's war experience and his relation to and growing distrust of art during his effort to record that experience in *Seven Pillars of Wisdom.* The third section turns to the later writings of the *Odyssey* translation and *The Mint,* drawing conclusions about why neither work represents, except ironically, the achievement of a poetic synthesis of self. And finally, the conclusion considers how Lawrence's end differs from Odysseus' own: one man reaches home while the other does not. Odyssean as it is, which is to say, as Odyssean as a modern life can be, Lawrence's experience finally diverges from that of the ancient hero and suggests profound implications for the possibility of heroic action in the modern era.

Odysseus' Poetic Integration of Self

In the scene leading up to Odysseus's dramatic triumph over the suitors, the disguised hero examines and strings his cherished bow: "Odysseus the master of craft had by now handled and surveyed the great bow up and down. Calmly he stretched it out with the effortless ease of a skilled musician who makes fast both ends of a piece of twined cat-gut and strains it to a new peg in his lyre. Changing the bow to his right hand he proved the string, which sang to his pluck, sharp like a swallow's cry" (21.404–11; 291). At the culminating moment of Odysseus' heroic return from obscurity Homer likens his hero to a poet. The comparison is not merely self-aggrandizing on Homer's part. Rather, it is a recognition of and a tribute to the fact that Odysseus' action is grounded in a poetic synthesis of experience. If it were not for his storytelling Odysseus would not have returned to Ithaca to stand on his palace threshold poised to shoot down the interlopers.

The poetic process that brought him to this point, however, developed over ten long years on top of the ten that he had spent in Troy. Setting out from that city, where he had fought valorously for the Achaeans, Odysseus enters a world where the *Iliad's* heroic ideal not only does not apply, but that imposes terrible consequences on those who attempt to follow it. The typical heroic act of proudly declaring his identity to the blinded Cyclops triggers the onset of troubles that eventually send Odysseus into prolonged exile on Calypso's island (9.502; 35). Again and again in the world of the adventures Odysseus indulges in futile Iliadic gestures. He draws his great sword only to cut his ship's cables and run, abandoning the rest of his fleet to cannibalistic destruction (10.126; 32); he draws it again to frighten a seductive goddess before sleeping with her (10.321; 47); and yet again to wave about impotently as Scylla snatches six victims from his already decimated crew (12.226; 46). In this strange world he repeatedly adopts unheroic postures of trickery and anonymity. He defeats the Cyclops, after all, only because he initially thought to call himself "Nobody" (9.364 ff.).

Along with enforced questioning of the heroic ideal, Odysseus must undergo increasing isolation over the course of the adventures. He sets out from Troy with twelve ships. After the early Laestrygonian disaster, that number is reduced to one. The crew's relations with their leader culminate in mutiny as they disobey Odysseus' command and devour the Sun's cattle on Thrinacia. No one can long defend the crew as a set of stalwart heroes, but neither can one argue successfully that Odysseus bears only minor responsibility for their fates. In part the antagonism underlying his relations with his men reflects the inherent antagonism between the heroic individual and the group. At the same time, however, isolation is a defining characteristic of Odysseus in particular. Described at key moments as "cut off" or "alone," Odysseus gradually sheds all his trappings—be they ships, men, war prizes, even clothes. This shedding of outer identity reflects in turn a deeper stripping of the self. The hero arrives on Calypso's island, "the navel of the sea" (1.50), haggard, defeated, and in a state of utter self-uncertainty. As is so often the case in Homer, external events express internal necessities: he stays there for seven years because he needs the protection of this goddess, whose name means to "conceal and preserve."

Odysseus's narrative in books 9 through 12 displays the poles of his own self-understanding. Throughout the adventures he plays widely varying roles, from the self-confident Iliadic hero to the lowly suppliant to the weak captain unable to ward off the evil impulses of his crew. No one of these roles fully reveals the "true" Odysseus, yet no one of them is merely spurious. His masks of humility and heroism represent throughout the poem two fundamental and fundamentally opposed aspects of his character. The poem consistently belies the implication of the hero's own tales that he is a scoundrel and a sloth, yet it also shows that the beggar disguise of the later books is not unwarranted. Together the conflicting poses describe his major inner conflict: nobody or hero? They also raise significant obstacles to the determination of the "authentic" Odysseus. It becomes difficult to say where a mask stops and the "real" Odysseus begins. Which comes first, the mask of humility or the humility itself? The mask of heroism or the heroism itself? The difficulty of pinning down the hero's authentic self belongs not only to us, but to him as well. Masks not only create ambiguity, they issue forth from it.

The problem of ambiguity comes to a crisis during Odysseus' visit among the Phaeacians, his last stop before reaching Ithaca at last. As a guest there he listens to three songs, two of which take his own role in the Trojan War for their subject. In response to both Trojan songs, the hero weeps. Clearly his memories of the war unsettle him; but he is also confronting for the first time his own image in art. Every hero of the *Iliad* longs for just what Odysseus here receives: *kleos* ("fame"), the celebration of his deeds in immortal song. But the traditional exchange for *kleos* is death. No Homeric hero has had to face his own immortality, whether directly or in song.[4]

Homer depicts the emotional intensity of Odysseus's unique situation in one of his most striking similes. As Demodocus sings of Odysseus's glory in the Trojan Horse ambush,

> Odysseus melted and tears from his eyelids bedewed his cheeks. So it is when a loving wife flings herself, wailing, about the body of her man who has fallen before his township and fellow-citizens, defending the town and his children from their cruel day of sack and rapine. The sight of him labouring his last breath and dying makes her wail aloud and wind herself about him. Yet do the enemy from behind beat her with their spear-shafts across her bowed shoulders and lead her into servitude, to her fate of toil and grief. Just as that woman's cheeks are ravaged with despair, just so piteously did the tears fall from Odysseus's brows. (8.521–31; 118)

The simile demands our attention not only by its comparison of the victorious Odysseus to a war-victim, but by its likening him in particular to a woman who has lost her husband. The sexual displacement suggests how deeply disturbing Odysseus finds the encounter with Demodocus' song: his very nature has been torn from him and he is left to view the world from an utterly new position, one of which he cannot have had any previous experience. Moreover, the position is one of imminent enslavement. The fame of immortal song turns out to unman and entrap its hero as much as it glorifies him.

Odysseus's identity is already tenuous when he arrives in Phaeacia: he has only just managed to leave Calypso after a seven year retreat from the world of men. He has been storm-beaten and washed up on shore, born anew from the ocean's belly into human society. Although he has begun to reestablish his heroic identity among the Phaeacians, his position remains an uncertain one, caught between life and death in the paradoxical space of poetry. Demodocus' song presumes him dead—or at best it is indifferent to his life or death—yet it perpetuates his immortal survival in a new form. Meanwhile, the living Odysseus sits listening to a song based on but independent of his life, a song that fosters not his actual life but a former life in which he can no longer play any part. Listening to himself represented in art, Odysseus learns one of the effects of his long absence: he no longer exercises exclusive control over his identity. The poet's re-creation of his life in song robs Odysseus of the ability to reconstitute his identity definitively. The life that he has accepted death to regain now risks enslavement to the rigid image and reifying power of art.

Odysseus, who has labored much and risked everything to reestablish his identity, learns before his journey is completed that no amount of activity on his part can enable him to regain absolute control over it. The realization disorients him and he weeps; but by the time the party returns to the palace for another round of feasting and song, he has decided how he will act. He sends Demodocus a prime cut of meat and his compliments, then praises the bard's skill and orderliness and asks for another song about himself (8.474; 98). Again he weeps upon hearing it, encountering himself again in a realm beyond his reach. But the very request to hear a song about himself offers the key to his response to his own mythologization: he will promote the telling of such stories in any way he can; he will contribute to his incarnation in myth. If he cannot own his identity through exclusive control of what others hear and say about him, he will at least make it as difficult as possible for them to hear and say truly.

Odysseus' policy of obfuscation begins in earnest with the tale of his adventures. His autonomy in telling these stories means that they might only be lies. Homer frequently calls everything the hero says into ironic question. We witness Odysseus tell what we know to be lies on a number of occasions and we note his great facility in doing so. If we take a prosaically realistic view of the stories he tells, we can pronounce them lies almost from the start, seeing them only as lively but insignificant fairytales. If, however, we poetically allow the truth of his tales to emerge from their subtextual levels, then we can begin to see how each of the adventures enables Odysseus to replay important psychological or spiritual issues. In emphasizing the Cyclopes' lack of laws and community, for example, and their absolute lack of the technical knowledge fostered by civilized life, Odysseus recasts in poetic form his climactic decision to leave the stultifying solitude of Calypso's island and return to the human community. In a poem as rife with wordplay as the *Odyssey,* the very similarity in the loners' names—*kalupso* and *kuklops*—hints that despite their radically different exteriors, the ultimate effect of isolation is savagery. Allowing truth to emerge poetically, then, we free ourselves to ask whether and how Odysseus' stories reflect inner truths and help him continue the journey of homecoming.

It shouldn't surprise us to find that one of these stories, the Sirens episode, replays precisely the hero's unsettling yet galvanizing experience of listening to Demodocus' songs in the Phaeacian court. There Odysseus describes how, thanks to Circe's detailed instructions, he has stopped his crewmen's ears with wax and is himself bound safely to the mast as the ship approaches the Sirens. "Come here, o come, *poluainos* Odysseus, great glory of the Achaeans," they beckon him.[5] *Poluainos* can mean "much-praised," which is how Lawrence takes it; or it can mean "the subject and/or teller of many *ainoi,*" which is how I take it.[6] According to Gregory Nagy, an *ainos* is "an allusive tale containing an ulterior purpose."[7] Nagy discusses the *ainos* in linguistic terms: it is a coded speech whose message will be apparent only to those who are able to understand it. In this sense the *ainos* is linked to riddling speech (*ainigma*) and to the riddling content of oracular response (*ainizomai*).[8] A hero's *philoi* ("dear ones") are the proper audience to understand an *ainos,* and such an audience may be created through narrative context. In the absence of *philoi,* the *ainos* imparts only its overt moral or narrative value; within a friendly context, however, the *ainos* acquires deeper significance for its audience.

The *Odyssey* as a whole provides the narrative context for Odysseus' *ainoi.* The poem offers to create an audience of *philoi* out of its readers, who might then go on to interpret and understand Odysseus' narratives. To the extent that Odysseus is perceived as telling lies—that is, lies insignificant for revealing the truth of their teller's self-conception—he merely covers himself over and might just as well reside still with the concealing Calypso. But however much the battered Odysseus needs and desires concealment, he cannot allow himself to remain completely hidden. The success of his return depends precisely upon his self-revelation and the recognition accorded to him by others. In calling Odysseus *poluainos,* the Sirens recognize both him and his need for the audience of *philoi.*

So begins the Sirens episode, which exemplifies the kind of *ainos* that its teller, within the *ainos,* is already famous for. In the *ainos,* Odysseus has just left Circe's island, where he had recuperated for a year from harrowing adventures; this departure from Circe's island corresponds to the hero's departure from Calypso's protective attentions in the outer frame of the narrative. Reaching the Sirens parallels the hero's arrival among the Phaeacians: in both places he encounters skilled singers whose songs threaten to enslave him, whether to actual death or to the death of inactivity. His report of being tied to the mast and then hearing the Sirens is a way of acknowledging his feeling of impotence when he hears Demodocus sing about him. The *ainos* reveals the ambivalence we have already seen: in both the inner and outer frames of the story Odysseus recognizes his powerlessness before the song of his own fame, yet he yearns to hear more, to lose himself in stories about himself, to hear without end the *kleos* that heroes were always promised. To stop and listen to the song, however, would preclude the possibility of acting in such a way as to achieve *kleos* in the future: the Sirens' song is dead, a purely retrospective view of a frozen past.[9] To stay with them means death, while to stay with the Phaeacians—a people self-proclaimedly distinguished by "eating and harp-playing and dancing and changes of clothes" (8.248–9; 110)—would entail the unheroic paralysis of idleness. In the *ainos,* tied to the mast, he has no choice but to go forward, just as on Scheria he knows he must pursue his original intention in leaving Calypso and not linger with Nausicaa. Odysseus' behavior with the Sirens thus parallels his behavior in Phaeacia: he admires the songs in both places, but in the end he must assert control over his identity and not leave it in the hands of others. He sails on in both stories, away from the Sirens and away from the Phaeacians.[10]

The temptation to bury oneself in an already established identity and past is a powerful one but would effectively put an end to the present and future. The cure to this temptation is not, however, to refuse the past altogether in the attempt to forget it—a course that Odysseus initially seems to adopt in his reluctance to talk about himself. Rather, the hero must confront the past through story and song. He must reconstitute it in a form that will aid his present and future life rather than hinder it. Odysseus, by remembering his past and reformulating it in story, thereby re-creates it. He gives it a form in which it may have meaning for him, rather than merely threaten him with its obliterating powers. The narration of books 9 through 12 constitutes an inward reliving and reinterpretation of experience.

In telling a tale that so remarkably recapitulates his experience, Odysseus seizes control over his identity and becomes a poet. He thus acquires his identity as a poet in order to preserve his identity as a hero. In doing so, he becomes much more than either a poet or a hero; each identity complements and deepens the other. Before he can recast his experience in the form of *ainos* Odysseus must think it through as both actor and storyteller. That rethinking of experience becomes central to the hero's character in the poem. Odysseus becomes a poet, a man of words, in order to interpret and take control of himself as a man of action. When at last he strings his bow as a poet strings his lyre, his poetic and heroic identities merge. The *Odyssey* thus posits an integral relationship between poetry and action: each grounds the other. Poetry reviews and comprehends action and so forms the basis from which heroic action becomes possible. Indeed, the *Odyssey* itself constitutes just such a review and comprehension of the *Iliad.*

In the *Odyssey,* the hero's relation to storytelling is the unifying element of his identity and experience. As a poet the hero embraces the tendency to isolation implied by his reflective nature while at the same time bringing himself into significant contact with the community through his stories and the need for recognition. By sharing the understanding of life that he has so painfully acquired, the hero benefits the community and attains a greater internal equilibrium than enforced privacy could offer. Such sharing thus enables him to embrace his human duties and relations, uniting his heroic action and identity with his reflective nature and experience. It enables him to live in the world of men.

Not all heroes are reflective after the manner of Odysseus. Achilles is not, and his experience illustrates the problems an otherwise outwardly turned man can have when the need and opportunity for introspection are thrust upon him. At the same time, not all heroes who reflect on their experience achieve the same unity in poetry. Like Odysseus, Lawrence tries to exorcise the ghost of his past experience through the poetic process, but in his case poetry seems instead to deepen his wounds. Writing *Seven Pillars of Wisdom* and *The Mint* and presenting them to others were necessities for Lawrence. But despite the brilliance of his prose and his fame as a war hero, these attempts to synthesize his experience aesthetically seem not to have brought him to an Ithaca of his own. The *Odyssey* teaches that poetry cannot exist in a vacuum, independent of life's obligations. Before Odysseus could achieve a living *kleos,* he had to return to life and accept its demands. Here we find an important difference between the fictional hero and the historical man. When the war ended, Lawrence accepted some duties but he also refused many. Above all, he had no personal obligations to resume in the way Odysseus did: he had no home to return to and restore. Poetry alone, in the absence of fulfilling duties, failed to heal Lawrence's psychic wounds. But it is time to turn to Lawrence himself, to examine the ambiguities of his experience and his attempt to comprehend life through art.

Lawrence's Attempt at Poetic Integration in *Seven Pillars of Wisdom*

One of the most fruitful of the many observations that link Lawrence to the ancient tradition comes from James Notopoulos, a classicist who approaches Lawrence through the metaphors of epic and tragic poetry. He likens Lawrence's active experience before and during World War I to the epic narratives of Homer, which he says help express the modern figure's "splendor as a human being and a writer."[11] But Lawrence's inner experience he assigns to the realm of tragic poetry. In the Englishman's frustration and suffering Notopoulos finds a second Philoctetes, "the unbending Homeric hero . . . made to face a new world."[12]

In the *Philoctetes*' conflict between modernity and "Homeric ramrodness," Sophocles cynically uses Odysseus to represent the modern merchant of values, adept at "diplomacy, compromise, [and the] use of people as means to an end."[13] Philoctetes, on the other hand, adheres to the older system, whose hero was Achilles. At the symbolic center of this debate stands Neoptolemos, Achilles' son. In his devotion to the heroic excellence epitomized by his father, Neoptolemos sympathizes with Philoctetes; but he arrives on Lemnos in the service of the Achaeans and as a follower of Odysseus, who seeks to capitalize on the youth's inexperience and uncertainty and to teach him the arts of deception. As Notopoulos notes:

> This confrontation is irreconcilable even for Sophocles himself, who in this play alone makes use of a *deus ex machina* to resolve the conflict. So too with Lawrence who carried in his soul a dialogue between the Homeric hero and a nonheroic world. Yet he could not find any *deus ex machina* except retreat into anonymity and the self-effacement of a recruit's life . . . The heroic bow unerring in its integrity to the heroic ideal has its price, the wound, both for Philoctetes and Lawrence. Both rage because of the wound, Philoctetes festering in the lonely isle of Lemnos, Lawrence in the Lemnos of his private self.[14]

Notopoulos aligns Lawrence with Philoctetes and by extension with Achilles in his desire for a system of absolute heroic values. Battles between past and present haunted Lawrence's spirit; he suffered "the dilemma of a modern figure who experienced the Homeric delirium of the brave . . . yet was condemned to tragic frustration by the anachronism of the heroic act in our times."[15] Lawrence himself, however, could never have lined up with Achilles. In *Seven Pillars of Wisdom* he confesses to a guilty role in the Arab revolt and lingers mercilessly on torturing thoughts of the betrayal of his followers. As he presents it, his relationship to the Arabs closely resembled Odysseus' relationship to Neoptolemos in Sophocles' play. Lawrence, the agent of modern Western imperialism, accuses himself of corrupting the youthfully innocent and noble Arabs much as Sophocles portrays Odysseus, representative of the new amorality, corrupting the young heir to Achilles.

Lawrence surely did long for the simple moral world of the heroic ethic as traditionally conceived. As one can see in Odysseus, however, that ethic, even in antiquity, was never as straightforward as the yearning modern imagination would like to believe. Odysseus's versatility results from his recognition of the conflict between a simplicity that cannot exist except in the imagination and a complexity that necessarily produces ambiguous moral values. Even the *Iliad* portrays its hero in the throes of despairing uncertainty as he faces the horrors of war and mortality, while the *Odyssey* asserts that no simplistic heroic ideal can possibly confront the intricate demands of peacetime life. We have seen that Odysseus himself would rather be flexible and ambiguous than fixed forever in time, petrified and immobile. Precisely by virtue of his deliberate ambiguity, which he exercises in the masks he wears and the stories he tells, Odysseus reappears in every new era, newly interpreted as its cogent expression of the "modern."[16]

Lawrence used ambiguity in the same way. He rivals Odysseus in suppleness and uncertainty of character and occupies a prominent place in the Odyssean procession of the modern types of past ages: in the eighty-five years since his campaign in Arabia he has transmuted from adored object of romantic longings, to hated imperialist spy and charlatan, to pitied victim of the divisive, alienating forces of the twentieth century, becoming at last but still not decisively the center of polytropic attention in collections of essays such as this one.[17] As Odysseus' before him, so Lawrence's life fosters repeated analysis and interpretation. The split sides of his reputation reflect, not fortuitously, the condition of Lawrence's inner life. He suffered from the modern cleavage of self that sets mind against body, motive against action, a cleavage that expresses itself through self-doubt and seems to render heroic action impossible—or, if possible, pathological—in our times. For Lawrence and Odysseus both, heroic life entails disorienting ambiguity of self, not only in later mythology but also in their own minds. Lawrence thus appears more Odyssean than Achillean, and the *Odyssey* provides not only a trope for Lawrence's "splendor as a human being and a writer" but also speaks to his "core of darkness"[18]—along with a great many other facets of Lawrence's many-faceted, even *polutropos* nature.[19]

Critics who view Lawrence in Homeric terms take their lead from their subject, who also saw the similarities between Odysseus' world and his own experience before and during the war: "For years we were digging up a city of roughly the Odysseus period. I have handled the weapons, armour, utensils of those times, explored their houses, planned their cities. I have hunted wild boars and watched lions, sailed the Aegean (and sailed ships), bent bows, lived with pastoral peoples, woven textiles, built boats and killed many men. So I have odd knowledges that qualify me to understand the *Odyssey*, and odd experiences that interpret it to me."[20] Lawrence understood the epic quality of the material that went into *Seven Pillars of Wisdom*. Accident, he writes, "with perverted humour, in casting me as a man of action had given me place in the Arab Revolt, a theme ready and epic to a direct eye and hand."[21] But, as he also acknowledged, his nature saw "all things in the mirror of itself, and not with a direct eye."[22] The passage from *Seven Pillars of Wisdom* continues: "The epic mode was alien to me, as to my generation. Memory gave me no clue to the heroic . . ." (*SPW* 549).

Lawrence's epic thus adopts a mode that he and his generation could understand. It is, as he realized, a peculiarly modern epic, an epic of introspection that inherently fails to be direct. From this contradiction arises much of the difficulty of the book, which offers at once a tale of epic character and action modeled on Homer and a searching, sometimes embarrassingly intimate exploration of the subjective self. These conflicting enterprises spawn a corresponding conflict between those readers who demand that the book maintain standards of historical objectivity and those who defend Lawrence's aims as fundamentally other than historical—a dispute reminiscent of that over the *Odyssey,* between those who would have it tell an exciting but largely superficial adventure story and those who would mine its depths for jewels of personal and human significance.

Responding to real and anticipated confusion about his intentions in *Seven Pillars of Wisdom* and the book's historical status, Lawrence expressed many conflicting and often directly contradictory views—views, moreover, that coexisted over time and do not reflect a progression in his thinking about the project as it moved forward. On the one hand, he insisted upon the book's literal historical accuracy. He had, he said, "a professional duty" as a trained historian "to record what happened."[23] The book "was forced from me not as a poem, but as a complete narrative of what actually happened in the Arab Revolt."[24] On the other hand, Lawrence understood very well the inherent untruth of storytelling, whether it takes the form of history-writing or treats its fictionality more self-consciously:

> One of the ominous signs of the time is that the public can no longer read history. The historian is retired into a shell to study the whole truth; which means that he learns to attach insensate importance to documents. The documents are liars. No man ever yet tried to write down the entire truth of any action in which he has been engaged. All narrative is parti pris. And to prefer an ancient written statement to the guiding of your instinct through the maze of related facts, is to encounter either banality or unreadableness. We know too much, and use too little knowledge.[25]

As a sophisticated reader of medieval chronicles, Lawrence had no expectation that accounts of events could escape the personal viewpoint of their chronicler—an expectation that applied equally to writers of wartime reports from the scenes of action. Nor was he likely to insist that truth be the determining factor of a history's value, writing at one point that "the prejudices of historians are generally the richest part of their histories."[26] Finally, completely aside from the question of how one should read and write history, Lawrence acknowledged that *Seven Pillars of Wisdom* was much more about himself than about the war and that it consequently embodied a peculiar kind of truth-telling. In the book itself he states that "in these pages the history is not of the Arab movement, but of me in it" (*SPW* 24). Later he wrote to E. M. Forster that the effort was "an orgy of exhibitionism" he would never repeat.[27] The personal content of the book resulted in Lawrence's deep ambivalence toward it, as he found himself characteristically torn, like Odysseus before him, between impulses of concealment and revelation.

When *Seven Pillars of Wisdom* is viewed as a primarily subjective document, factual accuracy becomes less an issue and a different kind of truth comes to the fore—as Lawrence understood:

> The personal revelations should be the key of the thing: and the personal chapter actually is the key, I fancy: only it's written in cypher. Partly it's a constitutional inability to think plainly, an inability which I pass off as metaphysics, and partly it's funk—or at least a feeling that on no account is it possible for me to think of giving myself quite away. There would be only two ways out of this—one to do like Pepys, and write it in cypher, as I have done—one to write what is not true, or not complete truth—and the second I don't like. . . . I think it's all spiritually true.[28]

Spiritual truth is equally the great aim of the *Odyssey,* and as we saw earlier the means to it lies in coded speech, or *ainos.* As Odysseus searches by means of his story to reconstruct a complex, living human identity, so Lawrence attempted through *Seven Pillars of Wisdom* to interpret his experience in morally intelligible terms. In the end it was as necessary for Lawrence as for Odysseus both to expose and conceal himself: he, too, could tell the truth only through fictions, and we must interpret Lawrence's narrative as we interpreted Odysseus's.

Odysseus and Lawrence perhaps share similar attitudes to truth and storytelling because they share similar motivations in speaking obliquely through poetry. Like Odysseus, Lawrence found himself during and after the war in a tangle of identities, uncertain not only of which one was real but of whether a real one existed at all. Where Odysseus had foundered in the gulf separating his identities as Iliadic hero and beggarly suppliant, Lawrence's uncertainty arose largely from the conflict, and its consequences, between Englishman and Arab. As one of its principal features, *Seven Pillars of Wisdom* portrays its hero caught in the whirlpool that results from these opposing torrents.

The nomadic Arabs of Lawrence's conception protected themselves against the desert's harsh expanse by narrowing the field of their vision. "They were a people of primary colours, or rather of black and white, who saw the world always in contour. They were a dogmatic people, despising doubt, our modern crown of thorns" (*SPW* 38). Lawrence knew that to lead the Arabs successfully he would have to become one of them, and he in fact attempted to imitate them so well "that they [would] spuriously imitate him back again" (*SPW* 31). But as a Westerner bred under the cool gray skies of the British Isles, he had little hope of truly achieving for himself the certainty fostered by the severe desert landscape. He admired the singleness of the Arabs as he admired the singleness of Achilles, yet he knew that neither condition was attainable for him. Even so, he strove to play out his role, which he maintained in part for political reasons, but in part also for personal ones—a confusion of motive that only aggravated the resulting confusion of self.

Despite its strains and dangers, Lawrence proved extremely successful in his pretense as an Arab. As he later suggests, "of course I must have had some tendency, some aptitude, for deceit, or I would not have deceived men so well . . ." (*SPW* 552). The success of his disguise, however, harrowed him throughout his Arab service and contributed to the mental distress from which he could not escape. Eventually he attributed to his prolonged disguise the destruction of his very self. He sets out this conflict in the opening pages of *Seven Pillars of Wisdom,* where he shows himself so deeply entangled in the dynamic of self and other that he fears madness:

> In my case, the effort for these years to live in the dress of Arabs, and to imitate their mental foundation, quitted me of my English self, and let me look at the West and its conventions with new eyes: they destroyed it all for me. At the same time I could not sincerely take on the Arab skin: it was an affectation only. Easily was a man made an infidel, but hardly might he be converted to another faith. . . . Sometimes these [active and detached] selves would converse in the void; and then madness was very near, as I believe it would be near the man who could see things through the veils at once of two customs, two educations, two environments. (*SPW* 31–32)

Faced with two conflicting identities and lacking the fixed point of reference that shored up the Arabs and certain other people he admired, Lawrence ended by losing all sense of identity.

Both Arab and British versions of identity represent aspects of Lawrence's war self—the self that, perhaps, could be heroic. But the two versions of self would be heroic in very different ways: the British taught a highly constrained, regulated kind of heroism, one that accommodates itself to armies and discipline; the Arabs were heroic individually, through a self-assertion that might at times act against the greater corporate body. Lawrence's affinities between these alternatives are difficult to determine. He saturated himself in chivalric tales of personal glory and achievement, even carrying Malory with him during the revolt. His letters suggest that he initially treated the revolt as a sublime adventure along such lines, and he presents in *Seven Pillars of Wisdom* the ideal of this kind of hero, Auda Abu Tayi, who descends upon them "like a knight-errant" (*SPW* 222). But although Lawrence includes figures like Auda in his book, by the time he sat down to write *Seven Pillars of Wisdom,* his belief in his own heroic possibilities had changed. He no longer accepted unquestioningly the value of the individual's self-assertion. As he would prove by enlisting in the RAF, he had come to embrace the group's mechanistic ideal of heroic achievement. This later, reduced vision of heroism infiltrates *Seven Pillars of Wisdom,* coexisting uncomfortably alongside the fabled Auda.

"Some of the evil of my tale," Lawrence begins, "may have been inherent in our circumstances" (*SPW* 29).

> As time went by our need to fight for the ideal increased to an unquestioning possession, riding with spur and rein over our doubts. Willy-nilly it became a faith. We had sold ourselves into its slavery, manacled ourselves together in its chain-gang, bowed ourselves to serve its holiness with all our good and ill content. The mentality of ordinary human slaves is terrible—they have lost the world—and we had surrendered, not body alone, but soul to the overmastering greed of victory. By our own act we were drained of morality, of volition, of responsibility, like dead leaves in the wind.

Lawrence's glorious dream-adventure turned quickly from a chivalric fantasy into the riveting nightmare these lines evoke. As he immersed himself in the Arab war effort, he found himself beckoned to act in ways violently contrary to the modern civilized ethic in which he grew up.[29] As the quoted passage suggests, he responded to the call and acted as required, but only at the expense of the inner code that had sustained him until then, as well as at the expense of his future mental balance. Stranded between his two worlds, he encountered a problem of action that would plague him throughout his life. Late in the book he admits that his moral position had become so eroded that he could scarcely bear to go under fire (*SPW* 502). By that point in the war his action had become too contrived, too difficult, too unnatural—in sum, too much cut off from its inner sources. As such, it could not be heroic. Like Odysseus, whose experience undermines the Iliadic ethic, Lawrence also comes to recognize the inadequacy of the deed. *Seven Pillars of Wisdom* reflects that realization and in this sense resembles the part of Odysseus' narrative that calls Iliadic heroism into question.

The war left Lawrence without the resources to say who he was. He was no longer "Briton," but had not become "Arab." He was not "hero." He knew only how hopelessly divided into parts he was. This problem of self had grown out of both his action and the reflection on action that followed it. The process parallels that of Odysseus, who also fights a war and finds himself afterward isolated with his thoughts, uncertain of his own identity and of his future actions. Odysseus' tale, as befits fiction, parcels out smoothly. Apart from human society for seven years, he has time to decide finally that he needs the recognition and community of other human beings. Perhaps more importantly, he has time to reach an understanding of his own human qualities, to formulate a conception of self that can withstand both recognition and non-recognition—that is, attempts by others to appropriate or to deny his separate existence. When he reenters the world to seek recognition, he has prepared himself. Lawrence's story, lacking that seven-year retreat, is less neat. His hope for an answer to the questions raised in him by action in war lay, as Odysseus' also had, in poetry rather than in further action. But Lawrence had less opportunity to reflect in advance about the power of art and what his art might look like. Perhaps as a result, he emerged from the London attic that witnessed "his decisive struggle with the angel"[30] not with a finally self-constructive tale, but with a document whose greatness lies, in part, precisely in the openness of the wounds it reveals.

Like Odysseus, Lawrence liked to tell stories, and one imagines him doing so with great relish and mischief—much as in his translation he portrays the beggar-hero in Ithaca telling tales about the beloved Odysseus' prowess. Lawrence's laughing presence is discernible behind the translation in such moments, as when the disguised king begs Penelope to restrain her tears, admitting, however, that "I cannot blame you, seeing how many women lament the dear dead fathers of their children, husbands not to be mentioned in the same breath with Odysseus, who all agree was godlike" (19.265–7; 262–3). Such mischievous self-promotion may have typified Lawrence's attitude when he first entertained the young journalist, Lowell Thomas, with wild stories about his exploits and those of his comrades—not realizing how they would later catch up with him in the form of an implacably importunate press.[31]

Again like Odysseus, however, Lawrence also told other kinds of stories: stories perhaps historically true, but perhaps not, stories that assumed spiritual importance as poetic syntheses of experience. The most significant story of this kind in *Seven Pillars of Wisdom* is the account of the Deraa incident, in which the disguised Lawrence was captured while spying in a Turkish city, severely beaten and perhaps raped. The discrepancies among his various accounts of the event, the critical controversy arising as a result, and the episode's importance as a defining moment in Lawrence's spiritual life warrant our looking at it as we looked at Odysseus's account of the Sirens: perhaps we may learn from it how Lawrence approaches the problem of self-integration through aesthetic reflection.

Lawrence intimates that at Deraa he gained terrible knowledge about himself: he learned there that he could derive sexual pleasure from physical pain:

> I remembered the corporal kicking me with his nailed boot to get me up. . . . I remembered smiling idly at him, for a delicious warmth, probably sexual, was swelling through me: and then that he flung up his arm and hacked with the full length of his whip into my groin. . . . Another slash followed. A roaring, and my eyes went black: while within me the core of life seemed to heave slowly up through the rending nerves, expelled from its body by this last indescribable pang. (*SPW* 445)

Characteristically, however, the *Seven Pillars of Wisdom* account is only one of many. Not to review in much detail here a controversy that has already felled more trees than it deserves, let it be said simply that Lawrence elsewhere hinted at a rape imperfectly resisted[32] and at the masochistic pleasure of desiring the very thing that most humiliated and tortured him.[33] At the same time, others of his writings give the event a different turn. Letters written immediately after the incident, for example, make light of it or give no indication that anything untoward had occurred.[34]

Critics have naturally raised many doubts about the truthfulness of Lawrence's narrative. Some say the incident never happened, others that it didn't happen in the way he describes, others remain undecided. Lawrence's practice, however, did not incline him toward wholesale inventions.[35] Deraa's importance for him—and ideally for us, insofar as we are his *philoi,* the created audience of friends to whom *ainoi* are addressed—was less historical than psychological, less material than spiritual. That he chose to represent whatever really did happen in the specific terms of the existing Deraa scene was a poetic decision and makes exquisite poetic sense. He felt his guilt toward the Arabs intensely and characterized his role among them as a prostitution. "Pray God," he writes in the first chapter, "that men reading the story will not, for love of the glamour of strangeness, go out to prostitute themselves and their talents in serving another race" (*SPW* 31). The Deraa scene is thus significant as Lawrence's self-presentation at that period of the revolt, and, more importantly, as his self-conception at the time of writing. Like Odysseus' narrative in books 9 through 12 of the *Odyssey,* the objective truth of Lawrence's tale cannot be determined. Like Odysseus, Lawrence left more than one version of aspects of his tale. Like Odysseus, Lawrence was a known storyteller. Finally, like Odysseus' stories, Lawrence's tales carry deep personal significance, representing in action the moral position he had asserted at length. The Deraa episode assigns to Lawrence in concrete, explicit terms the degradation he felt was his and the punishment he felt he deserved. In this horrific scene, thought and action are united. Deraa could not bear any greater weight of meaning if it *had* been invented.

Read in this way, the Deraa incident suggests that Lawrence's aesthetic interpretation of experience in *Seven Pillars of Wisdom* is finally more debilitating than constructive. Where Homer's hero succeeds in using art to cover over the scars of consciousness, Lawrence used it to deepen his wounds, perhaps even to inflict them. Odysseus' tales, like *Seven Pillars of Wisdom,* record the breakdown of the human self into anonymous uncertainty and its exile to the lonely reaches of a concealing island. But the narrative also discloses, and in fact constitutes, the sources of the hero's resilience. By telling his story Odysseus enacts his return to the human community. His poetry relates him to his surroundings and draws him back into the orbit of human life. Even for Odysseus, however, poetry alone was not sufficient to accomplish his return; first he had to have a home to return to. Lawrence's writing, if it failed to address itself to a welcoming human community, would also fail to enact his return.

One way to address a community is to offer it a political narrative, and according to André Malraux, Lawrence in fact began the book in order to make "an appeal to history against injustice."[36] But Malraux finds that the political narrative contradicts what the book ultimately became: "Little by little, without any change of frame, the Revolt began to take second place while the main interest passed to the absurdity of life for a man reduced to solitude by an irreducible inner conflict and the meditation it imposed."[37] What he had to write for the sake of the Arabs conflicted with what he had to write for his own sake. Malraux's Lawrence only gradually realized his truer, lyric purpose. Far from serving a political purpose, Lawrence's epic eventually became in his mind "the medium for a grandiose expression of human emptiness,"[38] an expression that would make its appeal to other men and be redeemed by the transfiguring lyric art of the book itself. But the book never achieved the desired transfiguration: "And the demon of the absurd appeared in the cruelest guise: if Lawrence had not expressed the man he believed he was, was it not simply because he was not that man? And if he was not that man, he was nothing."[39]

Malraux's Nietzschean conception of art may be overblown when applied to Lawrence, but it does capture an important aspect of the Englishman's poetic aspirations; it also recognizes, in highlighting the tension between Lawrence's personal and political aims, another version of the internal conflict we saw earlier between his longing for the absolute values of epic heroism as traditionally conceived and the more earthy realism of the practical Odysseus. Knowing better than to devote himself wholeheartedly to the former terms in these two oppositions, Lawrence treated his writing in precisely the wrong way for one whose aim was to achieve the kind of unifying vision of experience expressed in Malraux's transfiguring lyricism. Rather than plunge himself into his work like the Bacchanalian reveler into the abyss of Dionysian being, Lawrence approached self-expression in writing as he approached repairing a boat: first he sought the problem to express, and then looked to express it in the most appropriate terms. And the problem—himself—was elusive, so that the work involved in locating its proper expression was all the more difficult and the result all the more complex. He writes to Edward Garnett: "As you, a critic, have seen, the thing [*Seven Pillars of Wisdom*] is intensely sophisticated: built up of hints from other books, full of these echoes to enrich or side-track or repeat my motives. It's too elaborate and conscious a construction to admit simplicity—or rather, if I were limpid or direct anywhere people would (should) feel it a false stillness."[40] The creative picture Lawrence presents is one of gathering the nuts and bolts of his prose from various external sources and assembling them into a complex structure that he can tinker with indefinitely in the effort to adjust its levels of self-expression and self-concealment. Poetry to Lawrence is the mechanical effort of identifying, rather than creating, truth or value. Graves assessed the attitude toward art adopted by Lawrence in his later years:

> it was true, I now realised, that his interest in literature and art—in fact in everything not "practical"—had always been that of an outsider. He could, and did at times, write extremely well—when he confined himself to reporting the external aspect or mechanical workings of things. But he could not with any hope of success undertake work that required the exercise of fundamental values; because he did not believe in such values—he distrusted the mind.[41]

Lawrence looked to art not to create value out of randomness—the Nietzschean or Malrauvian project—but to express the concrete contents of his world. If nothing existed or if in looking inside himself he could not be sure what was there, then he had to struggle in his self-expression to be as accurate as possible in accounting for that lack.

Irving Howe sees a further purpose in Lawrence's treatment of prose as craftsmanship. *Seven Pillars of Wisdom* is "a work of the most artful self-consciousness," he writes, "in order to resist the pressures of memory."[42] As Lawrence delved deeper and deeper into memory, he required protection against the blinding glare of its truth. Treating representation and style as a matter of craft and logic provided just this protection. Jeffrey Meyers rightly quotes Nietzsche's aphorism—"We have Art in order that we may not perish from Truth"—in connection with *Seven Pillars of Wisdom.*[43] But it is important to see what "Art" here meant for Lawrence: it was not so much the ecstatic creation of beauty as veiling illusion envisioned by Nietzsche and embodied, he thought, by the Greeks. Lawrence resorted rather to the formal, craftsmanlike aspect of art, which protects even as it exposes. He once accused Homer in the *Odyssey* of trying "by surpassing pains and skill to simulate the rule-leaping flood of authentic greatness . . . He was out to construct an epic, on the *Iliad*'s model: and it all smells forced."[44] Lawrence's own surpassing pains and skill tended in a similar direction and, in his view, ended in a similar failure.

Lawrence denied setting out to construct an epic from the admittedly epic material given him by the Arab Revolt.[45] But the denials arose more from his later sense of failure than from an initial lack of intention or hope, as his ambition to make *Seven Pillars of Wisdom* an English exemplar of spiritual sublimity in literature amply shows. He did not, then, begin *Seven Pillars of Wisdom* with the technical approach to art I have been describing. Thomas O'Donnell argues at length that the debilitating introspective aspect of Lawrence's project gained prominence only in later drafts as the moral problem of his role in the revolt gradually took over his thoughts, finally consuming them and driving Lawrence himself to the brink of madness.[46] On this view, art not only failed to save Lawrence from truth, it drove him on relentlessly in search of it and ultimately destroyed him with it. He responded by taking an increasingly obsessive interest in the formal aspects of the book: the printing and typography, the illustrations and portraits, and the "mechanism" of the prose. Over a period of four years he attempted to heal, by picking at, the wounds left by the initial fiery search for truth. During that time he revised again and again the text of the book, which he had effectively completed in 1922.

Malraux may thus be right in attributing to Lawrence the ardent desire to achieve a transcendent vision of self-unity through art. But it was a flame that faltered and died out as work on *Seven Pillars of Wisdom* progressed. In Malraux's view the profound artist thrives not by technique and assertion, not by proficiency of talent, but by the passion and grandeur of his vision, by the sublimity of his spirit and his longing to immerse that spirit in the "Absolute." Lawrence finally stood too divided as a man, and too unwilling to overcome that division through self-immersion in the Absolute, to realize such a unifying truth in poetry. As one who "experiences the anguish of being a man to the depths of his being," he ranked in Malraux's eyes among "the most religious spirits of his time" (531). But he did not achieve the self-identification of the great personality, who "exists precisely in the bond between thought and act" (530). Lawrence balked at finally letting himself go: for all his self-repugnance, he clung desperately to the only self he knew. He thus seemed on the one hand to accept the Malrauvian construction of the problem of self, which envisions self-loss through immersion in the Nietzschean chasm of being. But on the other hand the all-consuming nature of

Dionysian ecstasy repelled him even as it attracted him, and his refusal of it left him without a totality, without an Absolute that he could wholeheartedly embrace. His religious longings had no God, no external, ideal standard.

Perhaps Lawrence feared the revelation that belief in self-transcendence through art might have brought, something he implies at one point in *Seven Pillars of Wisdom,* when he stops to bathe in a beautiful pool among the stupendous hills of Rumm. There he meets a mysterious old man who speaks to him of love and God. The speech disturbs the Englishman's doubts with an intimation of truth. One critic, referring to Forster's *Passage to India,* calls this scene "Lawrence's Marabar Caves experience; but where Mrs. Moore learns in the caves that nihilism is the truth of the universe instead of Christianity's doctrine of all-embracing order and love, Lawrence here glimpses love and fertility amid the meaninglessness and confusion."[47] But Lawrence rejects the lesson and refuses baptism. "In fear of a revelation, I put an end to my bath, and advanced to recover my clothes" (*SPW* 357). His religion will remain private and individual. As such, not even the power of art can rescue it with a saving vision.

Historically true or not, the episodes of *Seven Pillars of Wisdom* show how Lawrence attempted to use poetry for life, to resolve problems through art, and to reveal himself to the proper audience while concealing himself from hostile readers. Critics like Richard Aldington resemble Penelope's suitors, who suffer gravely from their refusal to interpret signs. The suitors laugh hysterically at prophecies of doom; like them, Aldington points a finger and says, "He's lying." No doubt Aldington and readers like him will not end up in Hades prematurely, but neither will they understand their subject. For his part, Lawrence would have welcomed this state of affairs. To the degree all interpreters must mistake their subjects he was content: "as long as it isn't true, I don't care what praise or blame I get."[48] To Graves he wrote, "I'd wish every man to be an everlasting question-mark."[49]

Lawrence has proved a particularly difficult question mark to penetrate, in part for the reason, which also applied to Odysseus, that he refused to cooperate in discussions whose focus is single-paned—a fate that is admittedly hard to avoid. Based on Lawrence's self-presentations as he meditated on his moral involvement in the revolt or on the events in Deraa, and based also on the analyses of readers such as Malraux, one comes away with an image of the man as an inveterate metaphysical brooder. So, too, one risks misreading Odysseus as a proto-philosopher-king, completely thoughtful and solemn. Seeing Lawrence and Odysseus in such lights does not wholly violate their natures. Both heroes and the works that bring them fame legitimately inspire metaphysical discourse. Lawrence really was obsessed with the problems of integrity and identity and the anguish of human existence, just as Odysseus really is a soberly reflective hero of tragic knowledge. On the other hand, one critic makes a fair if slightly overstated point when he observes that the "term 'transfiguration,' irrelevant to our understanding of the rather non-spiritual and earthbound Lawrence, is perfectly appropriate to our understanding of Malraux's deepest ambitions."[50] So too has Odysseus been fairly seen, including by Lawrence himself, as an uninspiring figure, clever and tough and susceptible of allegorical readings but in the end not very interesting metaphysically.

The mistake made by interpreters on both sides of this issue arises from their shared expectation that grandeur in human life cannot coexist with the mundane and even silly realities of daily existence. As a result, their heroes are either perfectly sublime, whether in aspiration or in achievement, or they are seen through, shown up, and exposed as frauds. Depending on the temperament of the critic, Lawrence and Odysseus have been placed at both ends of this heroic spectrum. In judgment, Lawrence himself sometimes subscribes to this either-or view of the hero. The mere existence of self-doubt bitterly entailed, in his eyes, the failure of his heroic dreams; and he explicitly condemns "that cold-blooded egotist Odysseus" as a heroic fraud.[51] In practice, however, Lawrence's affinities lay more with Odysseus the egotistical prankster than with the unflaggingly serious hero of popular conception. Those affinities emerge implicitly in his translation of the *Odyssey,* parts of which clearly suggest the translator's fondness for the hero in his best moments of self-ingratiating blandishment. They emerge explicitly as well, not only in his actions and writings, but also in the unanimous voice of his friends in recollecting his sense of humor. Lawrence did not pursue a humorless life for the sake of grandeur. Well-known as a practical joker, he typically targeted the excessive solemnity characteristic of just such affectation. *Seven Pillars of Wisdom* itself is full of humor. One of the funniest moments in that book occurs at the expense of its most epic character, Auda, whom Lawrence lampoons by imitating his epic style of descriptive action (*SPW* 278–79). For all his tragic knowledge and exalted spiritual aspirations, the Englishman, like the Ithacan, stood firmly rooted to the earth and its daily realities.

In resorting to earthly life, Lawrence resembles Odysseus. But in regarding the life of the flesh as at the same time shameful and corrupt, he parts company with his precursor. The difference of world-view separating the ancient and modern men shows up in their use of poetry as well. Odysseus's ultimately constructive, creative stories had enabled him to comprehend his past experience and to transform his identity without abandoning his individuality to some higher or profounder entity. He used the creative powers of poetry to his own advantage while accepting the corollary risks, represented in the Sirens episode, of immersion in the past and the loss of self-possession. But Lawrence shrank from the power of poetic creation. Unlike Odysseus' stories, as also unlike the Nietzschean artist of Malraux's conception, his self-invention negated the value of self and life. His self-presentations in art led not to affirmation of life but to withdrawal from it. They led to his attempt to live out in the ranks the degradation he had written into *Seven Pillars of Wisdom.*

In the end, then, although Lawrence rejected the romanticization of art, he did not fall back on a more truly Odyssean approach, one that would embrace art's creative powers without succumbing to self-abnegation, and that would accommodate both humor and greatness in its vision of life. In finally rejecting sublimity, Lawrence resorted not to this Homeric conception of poetry but to the safety of a mechanical ideal. Instead of pursuing the earthy, playful, and finally constructive model of poetry offered by Odysseus and Homer, Lawrence took refuge in a mechanistic approach to creation. He thus turned from the fear of self-loss through heroic and artistic sublimity into the arms of the machine—the ultimate non-self.

The Mechanical Ideal in Life and Art

In a letter to Robert Graves written very near the end of his life, Lawrence explained how he had changed since they had known each other at Oxford: "I was then trying to write; to be perhaps an artist. . . . Well, I failed in that. . . . Almost I could be an artist, but there is a core that puts on the brake . . . So I changed direction, right, and went into the RAF . . . to serve a mechanical purpose, not as leader but as a cog of the machine. The keyword, I think, is machine . . ."[52] But the RAF was not the sole form assumed by Lawrence's mechanical aspiration, which also resulted in two important literary activities: translation—in particular, his translation of the *Odyssey*—and writing what he considered a photographically observed account of life in the ranks, *The Mint.* In light of our interest in the encounters between Lawrence and Odysseus, we will focus on the first of these activities as our way of determining whether Lawrence's plunge into the other was finally successful. If it was, it would amount to a homecoming of sorts—to be sure, not a return to wife and family but a heroic renunciation of self and freedom from the onus of individuality.

The Dynamic Between Self and Other in the Odyssey *Translation*

Translation was a way for Lawrence to live with words and ideas without assuming the burden of creation. "In translating you get all the craftsman's fuss of playing with words, without the artist's responsibility of their design & meaning,"[53] he wrote, having long ago decided that "it is not good for a man to make things."[54] For Lawrence, translation was a safe outlet for his pleasure in word-crafting, because he thought that the self was inessential to the works he translated. He did in fact translate two lesser books swiftly and successfully in accordance with this model, and even after completing the *Odyssey* he complained that the translation did not have "anything of mine in it": its "second-hand English is no more personal than the filling in of an income-tax form."[55] His belief that he was not personally implicated in the translation may have freed him to criticize the poem and his version as if he stood far removed from them. If he was strictly faithful to Homer, he could legitimately claim to be free of all responsibility for his rendering, good or bad. If his translation was a failure it was not because he failed as a translator, but because Homer was "an antiquarian, a tame-cat, a book-worm: not a great poet, but a most charming novelist. A Thornton Wilder of his time. My version, and every version, is inevitably small."[56] All he could do, as he said, was "expose the fraud."[57] The Translator's Note conveys a similar impression. The *Odyssey,* he says, is "neat, close-knit, artful, and various: as nearly word-perfect as midnight oil and pumice can effect." Its poet, he believes, was conscious of the failure of his grand effects, of the "limitation of the work's scope," for he "lived too long after the heroic age to feel assured and large." Even so, the poet "shows exact knowledge of what he could and could not do. Only through such superb self-criticism can talent rank beside inspiration."

Lawrence's criticisms of Homer, however, apply more precisely to his own creative efforts and fears than to the poet of the *Odyssey.*[58] In *Seven Pillars of Wisdom,* Lawrence had aimed at the greatness of epic, just as the *Odyssey*'s poet, in his view, had set sights on the *Iliad.* Such success as *Seven Pillars of Wisdom* achieved resulted from Lawrence's emotional engagement in its subject. In the end, however, he too lived too far distant from the heroic age to feel "assured and large." As we have seen, circumstances had placed him in the midst of an action that invited epic treatment. He sought to turn his epic into literature, but found that "the epic mode was alien to me, as to my generation. Memory gave me no clue to the heroic . . ." (*SPW* 549). Lawrence knew his place in history, he knew exactly "what he could and could not do. Only through such superb self-criticism can talent rank beside inspiration." His criticisms of the *Odyssey* are his own self-criticisms; he accused Homer and Odysseus of the very smallness that he feared to find in himself, as artist and as man of action.

Views of Homer that seemed simply impatient or irreverent take on new significance when seen as symptoms of an encounter that turned out to be far from merely mechanical. We are led to wonder what role, after all, Lawrence's personality plays in his translation and assessments of the *Odyssey* and its hero. The Translator's Note attests to Lawrence's self-assertion in its long speculation on Homer's background, his personality, his habits and knowledge, his temporal relation to the heroic period, his language, and so on. But the best evidence of Lawrence's self-involvement is the translation itself, which reveals the translator in close competition with his original author. Lawrence's treatment of individual Greek words, for example, often manages to respect their integrity while rendering them idiosyncratically. For instance, he translates *thaumazen d'Odusea en ophthalmoisin horosa* (literally, "she wondered at Odysseus, seeing him in her eyes") as "admiring Odysseus with all her eyes" (8.459; 116). The phrase *en ophthalmoisin* is a characteristic feature of Homeric language, an archaic locating of the place in which a mental or physical activity occurs. Such linguistic peculiarities, so foreign to modern English, are typically omitted from contemporary translations. In rendering the words, Lawrence maintains strict literality; but in altering them to "with all her eyes," he tweaks them into a smooth, even colloquial, phrasing. Writing of *Seven Pillars of Wisdom,* Howe notices that Lawrence

> often uses words with a deliberate obliqueness or off-meaning, so as to charge them with strangeness and potential life. The common meaning of these key words is neither fully respected nor wholly violated; but twisted, sometimes into freshness and sometimes into mere oddity. All of this followed from a conscious planning: "I find that my fifth writing . . . of a sentence makes it more shapely, pithier, stranger than it was. Without that twist of strangeness no one would feel an individuality, a differentness, behind the phrase.[59]

Howe's observations apply equally well to the *Odyssey.* Lawrence's digging beneath the surface of the Greek words to resurrect their usually ignored meanings in unconventional and slightly non-literal renderings illustrates his mental attitude toward the fullness of Homer's language and his obligations as translator: on the one hand he envisions the scene more vividly and adheres more strictly to the literal sense of the words than do other translators; on the other hand, and in the very same moment, he asserts individual ownership of that vision. Lawrence thus strives simultaneously for literal fidelity to Homer and for creative independence from him.

Other aspects of the translation similarly reflect the intense dynamic between the translator and the ancient poet. Lawrence's treatment of Homer's repeated epithets and phrases, for instance, reveals especially well the competitive nature of his relation to the original, as if he sought to outdo Homer by refusing to resort to uniform, repeated, and therefore unimaginative phrasings. To take just two examples: Homer uses the phrase *epea pteroenta* ("winged words") sixty-three times in the poem. Lawrence, incredibly, renders it fifty different ways, omitting it altogether only six times and exactly repeating himself only twice. A more complex instance arises in Lawrence's treatment of the Homeric "rosy-fingered dawn" formula. For the twenty occurrences of the phrase *emos d'erigeneia phane rhododaktylos Eos* ("when early-born rosy-fingered Dawn appeared") Lawrence offers nineteen different renderings. Initially, he aspires to such poetic touches as, "So soon as rosy-fingered morning came forth from the first grey dawn," or "Day-break: and the rosy-tinted fingers of dawn crept up the sky." Soon the phrase, while still fully translated, becomes slightly less expansive: "At dawn's first redness in the sky," or "At the first red finger of daybreak in the east." In the end, however, Lawrence settles into what seems an increasingly irritated posture as he goes on to render the phrase in such clipped tones as "At the first show of Dawn," or "Dawn. When She came, rosy-fingered," or even simply, "At dawn."

This progression from the expansively poetic to the compressed, functional treatment of Homer's verse—a trend evident in the translation in many other cases than the one given above—implies a change in Lawrence's approach to translating the poem, as if he began to think that he had challenged Homer on the wrong (poetic) basis, and consequently altered his own style in accordance with his new understanding of Homer's. Thus the translator takes on in English the more naturally compact syntax of the Greek, and the competition continues on a new footing. It turns out that translation, at least of the *Odyssey*, was not mere wordplay for Lawrence. He could no more maintain perfect humility in this work than he could in writing *Seven Pillars of Wisdom*. Regardless of his own comments, which anyway he modified in later years,[60] the translation itself backs up the claim that in the end the translator "lacked the surrender of soul to submit himself lowly and reverently, even to the first poet."[61]

However successful a translator's immersion in Homer, the result is necessarily distilled through the lens of his personal vision, in greater or lesser degree. Homer is too far away to be got pure, and the translator must invoke imaginative freedom in order to approach Homer's distant world at all. But he must also, insofar as he aspires to Homer, subordinate his freedom and imagination to Homer's genius when once he locates it. This paradox of self-suppression and self-expression, which characterizes the act of translation, also and with remarkable felicity characterizes Lawrence not only as translator but as writer in his own right and as man of action. The paradox bears a striking resemblance to his earlier, more painful wartime dilemma of selfhood:

> A man who gives himself to be a possession of aliens leads a Yahoo life, having bartered his soul to a brute-master. He is not of them. He may stand against them, persuade himself of a mission, batter and twist them into something which they, of their own accord, would not have been. Then he is exploiting his old environment to press them out of theirs. Or, after my model, he may imitate them so well that they spuriously imitate him back again.

> Then he is giving away his own environment: pretending to theirs; and pretences are hollow, worthless things. In neither case does he do a thing of himself, nor a thing so clean as to be his own . . . (*SPW* 31–32)

Translating Homer presents similar difficulties, however smaller in scale. The translator may apply himself and his environment to Homer, but then he presses Homer out of his own world; or he may try to learn Homer by self-saturation, imitating him so well that Homer's voice begins to take on a ring of the translator's own and the two become confused in each other. In neither case is Homer speaking directly or purely through the translator; in neither case is the translator doing "a thing of himself" or "a thing so clean as to be his own." In Arabia Lawrence chose, or succumbed to, the ideal of perfect imitation, with the result that he lost his native self without retrieving a new, stable self in return—he was left, as he says, with "affectation only." Translation, while not as threatening to Lawrence's selfhood as his war experience, posed similar problems for the relationship between his native thought and idiom and the distant, alien force of the original authorial voice. The skill with which Lawrence imagined himself into Homer's scenes and then reemerged to paint them so vividly for us implies an experienced adeptness in moving back and forth between vastly different, foreign spheres—an adeptness, perhaps, born only from painful uncertainty of one's own authenticity.

The paradox of selfhood that plagued Lawrence, both as translator and as author of *Seven Pillars of Wisdom,* also besets Homer's hero. One wonders to what extent Lawrence recognized himself in Odysseus's experience. The proem, with its solemn evocation of the man "who ached in an agony to redeem himself," suggests that Lawrence will explore the depths of the hero's character. But the translation, while presenting many sides of the "various-minded man," seems ultimately not to accord the hero a profound nature. Perhaps Odysseus' experiences reminded Lawrence too painfully of his own: when Odysseus once "punished himself with humiliating stripes and threw a coarse wrap about his shoulders as if he were a bondman: and so went down into the broad streets of the hostile city amongst his enemies, hiding himself in foreign shape and making believe he was a mendicant"(4.244–48; 51)—perhaps in translating these lines Lawrence thought of Deraa, and rendered *plegesin aeikeliesi damassas* (literally, "subjecting himself to disfiguring blows") as "punished himself with humiliating stripes," because punishment and humiliation were what preoccupied Lawrence himself in that episode. Perhaps, too, Odysseus' struggles with selfhood—the conflict between his desire to be known, to proclaim his identity to the world, and his long if not unpleasant durance first in Calypso's "blind silence" and later in self-effacing disguise on Ithaca—rang too hauntingly in the ears of the translator, who insisted on retiring anonymity, on being a cog in the machine, but kept revealing himself as if under compulsion, both in the ranks and as a writer. Perhaps, finally, the *Odyssey*'s "tension—'ultra-modern' if such an epithet has any meaning—between the survival, the potential immortality, of the hero when he becomes the object of the minstrel's art, and the concomitant eradication of this same hero from actual life"[62]—with which it "vibrates," stung Lawrence deeply, whose own life was the subject of fictions and mythologies.

These conjectures may go too far beyond the more common observations regarding Lawrence's unusually Homeric experience before and during the war among the highly individual, honor-sensitive Bedouin. But they are not unwarranted. A reader of Pope's *Iliad* writes that for purposes of credibility, the translator of Homer "must have a *mediate term* in mind between Homer and us. He must believe in a hero who at the same time appears like a modern hero and like what Homer writes about."[63] Lawrence's "mediate term" between Homer and us may have been himself, with the consequence that he could not present Odysseus as finally heroic. If he did identify with Odysseus, it would redound to the Homeric hero's disadvantage: no one whom Lawrence perceived to be like himself survived intact from his criticism. The translator's developing censure of Homer and Odysseus parallels the progress of his deepening involvement in the translation and in itself suggests that just such an identification had occurred. In any event, the *Odyssey* was Lawrence's last translation job.

The Mint *and Self-Integration*

Our look at Lawrence's *Odyssey* has suggested that even after abandoning the artist's creative effort for the merely mechanical play with words, Lawrence found himself involved in a dynamic of self disturbingly like what he had suffered in writing *Seven Pillars of Wisdom.* Lawrence's other principal literary project during this period was a work very different from any he had yet undertaken, a work that, on the one hand, did not enable its author to overcome self-division and self-uncertainty through art, but that, on the other hand, does amount to a coherent expression of self. Although we cannot here read *The Mint* in detail, some reflections on it are in order before drawing our inquiry to a close.

In the fall of 1922, Lawrence enlisted in the Royal Air Force. He had served fewer than six months when the press discovered and broadcast his whereabouts, leading to a temporary expulsion from the force. In those months, however, he had begun to compile notes of his experience, which he later described as photographic excerpts of daily life.[64] And despite its hard degradation, he had begun to feel he had found a place at last, perhaps even a home: "And me? I had shrunk into a ball and squatted, hands over face, crying babily (the first time for years) on one corner of the scudding lorry, which rattled like a running skeleton, and at each leap dinted the impression of one projecting bolt or other into my substance. I was trying to think, if I was happy, why I was happy, and what was this overwhelming sense upon me of having got home, at last, after an interminable journey" (*M* 119). Lawrence returned to the RAF in August 1925. Once there he resumed his notes, editing them into *The Mint* during his posting to India between 1927 and 1929.

Most of the book describes the enlistment and training of new recruits—the process that takes "The Raw Material" (part one) and puts it "In the Mill" (part two). Lawrence recounts with uncompromising harshness the often pointless cruelty of the discipline and the simple vulgarity of the men. He illustrates unrelentingly the depth of the plunge he has taken into the muck of physical reality, which had always both repelled and intrigued him. This effort to immerse himself in what he most fears, to become like what he most distrusts, marks Lawrence's attempt to carry through to conclusion the debasing implications of his self-presentation in *Seven Pillars of Wisdom.* By devoting himself to physical existence alone, he hopes to locate in an exclusively earthly environment the core of self that had eluded him in the war. So far from keeping him "mask-less," his previous existence "had built me such a caddis-shell as almost prompted me to forget the true shape of the worm inside. So I had sloughed them and it right off—every comfort and possession—to plunge crudely amongst crude men and find myself" (*M* 27).

The "worm inside," however, turns out not to resemble very closely his comrades in the RAF. His literary attention to the minting process in itself highlights his isolation and difference from the men. He had undertaken the effort of service, he says, "partly to replace myself in a world from which much solitary thinking had estranged me" (*M* 60). But if he offers up his body to this process of relation, he resists the mill's attempt to co-opt his mind. No amount of flat assertion of unity with the men—an assertion Lawrence often makes through his use of the pronoun "we"—can overcome the separation between them.[65] Lawrence can learn a solidarity of sorts, but he understands that he will never actually be like the men, and that "I shall never be quite happy, with the happiness of these fellows who find their nectar of life, and its elixir, in the deep stirring of some seminal gland" (*M* 195).

Despite his difference, Lawrence nevertheless pursues integration, both between himself and the community and between the opposed parts of the self. These conflicts between individual and community and between flesh and spirit saturate *The Mint,* but always with an expressed hope of being overcome through integration. For this reason, Lawrence does not present RAF life as merely degrading, for he also thinks of his service as possibly ennobling or redeeming. He would join the group, submit to the machine; but he would emerge an individual, renewed and internally supported by his association with the community. This hope is the burden of the book, which its third and final part, "Service," attempts to realize. Its scenes derive from Lawrence's later experience, when he was posted to the Cadet College in Cranwell. The brutality of training had come to an end, and discipline was not so harshly enforced. The section is happier, calmer, and claims for Lawrence a newfound peace of mind as a common airman. It claims to heal in moments of integration the violent splits within the self and between self and world. Towards the end of part two, Lawrence anticipates this integration:

> Man, who was born as one, breaks into little prisms when he thinks: but if he passes through thought into despair, or comprehension, he again achieves some momentary onenesses with himself. And not only that. He can achieve a oneness of himself with his fellows: and of them with the stocks and stones of his universe: and of all the universes with the illusory everything (if he be positive) or with the illusory nothing (if he be nihilist) according as the digestive complexion of his soul be dark or fair. Saint and sinner touch—as great saints and great sinners. (*M* 179)

The final passage of the book asserts just such a moment of oneness. Lawrence describes himself lying on his back amid a heap of lazing bodies, at one with the earth and his fellow man, absorbed in thoughtless contentment (*M* 249–50). Critics divide on the authenticity of Lawrence's experience in this passage. On the one hand. Lawrence does seem to have entered a new stage of life, a stage free of the emotional traumas that characterize the *Seven Pillars of Wisdom* period. The RAF anchored him to the world and enabled him to forget himself often enough to become occasionally happy. But it did not instill in him an anchor of his own, nor did it help him to become self-sufficient in happiness by acquiring an internal basis for the harmony he seeks. In the end, then, he still had nothing on which to depend except himself. On February 25, 1935, his last day in the service, he wrote to Ronald Storrs: "I leave here tomorrow a.m. . . . and the R.A.F. that same moment. ειθε δε μηδ' . . . [66] The Greek indicates a contrary-to-fact wish—"if only not . . ."—by which Lawrence may simply have meant that he wished not to leave the RAF. But the words come from a famous epitaph, which Storrs provides in full: ενθαδε κει μαι / Ταρσευς μη γημας ειθε δε μηδ'ο πατηρ ("Here I lie, / Of Tarsus; never married; and would that my father had not.") Standing once again before the abyss, Lawrence wished—not for the first time—that his parents had not borne children.[67]

Granting, however, that Lawrence sometimes experienced the momentary onenesses of absorption described in *The Mint* while also granting that he did not live out such absorption fully, our task becomes to examine the nature of the integration he did achieve. *The Mint* shows that Lawrence found peace only in moments of wholly unthinking, unself-conscious physical being. Although not marked by the metaphysical exultation implied in Malraux's vision of submergence in the Absolute, Lawrence's absorption involved a similar abandonment of individuality and self-consciousness. But such moments do not give birth to books, which can only, as *The Mint* does, describe the absorption in retrospect or imagination. Lawrence expressed again his view of the irredeemable separation of thought and oneness when he wrote to Forster that he no longer took notes on the RAF: "I'm it, now."[68] The comment implies the cessation of self-consciousness in his service life. Although his letters show that this implication remains unfulfilled—Lawrence never abandoned his self-consciousness—the claim makes his point: if he went on writing about the RAF he would never become a part of it, and he wanted to become a part of it, a happily functioning "cog of the machine." However much poetry may help Lawrence present and examine a problem, it did not itself constitute the solution and in fact impeded it as thought impedes oneness. If Lawrence had the sense to reject the romantic conception of the sublime artist, neither could he accept Odysseus' actual solution and make poetry the ground of the unified self. His poetry remained disintegrating only: he used it to break himself down, not to weave himself together. If he then composed a scene of unity, as he did in the last paragraphs of *The Mint,* the scene remains an assertion, a representation of something that happened elsewhere, outside the book.

The scene of unity remains an assertion in another respect as well. For Lawrence imbued the RAF with the qualities of a religion before whose God the airmen must abase themselves in pursuit of personal redemption.[69] *The Mint* reverberates with Christian language: "Just as the roomy, sordid, clanging, momentous hangar is our cathedral, so our day's work in it is worship."[70] Lawrence comes to see human domination of the air as a faith; he calls the RAF a "lay monastery" and considers his participation in it as a spiritual quest that links him to other men.[71] But the "we" in such passages strikes the reader as especially unwarranted, and Lawrence himself betrays a creeping recognition that his comrades may not worship with him at the same altar, when he suspects that the airmen of the future "will not be so owned, body and soul, by their service" (*M* 236). One critic recognizes both Lawrence's longings as an apostle of the new religion and the inevitable insufficiency of this quirky faith as spiritual sustenance for other men. He writes that *The Mint* is "an essay in moral immolation and intellectual asceticism, religious in prompting, escapist in enactment; so that its final significance, while religious, is the significance of an irrational, restricted religion; which is to say that it is abnormal in the sense of being *merely* individual."[72]

The Mint, although a very different book from its magnificent predecessor, thus founders on the same rock of radical subjectivity as *Seven Pillars of Wisdom* had done; and Lawrence again failed to realize a meaningful individuality through transcendence of his isolation. In devoting himself to a religion of his own invention, he did not really give up his individuality at all. The desired community with others remained elusive. He had not built into himself an external standard shared by others but had rather incorporated those others into a vision he alone had constructed. As he discovered in leaving the RAF, that vision could not sustain him in separation from the physical reality in which his imagined construction flourished.

The paradox remains, however, that Lawrence could have found no better expression for his *whole* self than this enlistment in the RAF, where he could feel at once degraded and superior in accordance with the conflicting impulses of his self-understanding. The enlistment was truly an act emanating from self; but since this self was divided, it necessarily fell short of achieving the stated goal of making Lawrence conform to an integrated pattern. He could insist upon his degradation without falseness; yet he could not avoid his mental separation from the other men. Like other actions that ostensibly argue his humiliation, the enlistment in fact carries an ambiguous meaning. In every case Lawrence was both in control and at the mercy of others. Insofar as this dual situation accurately reflected his self-understanding, his action in the world did in fact express self. Since the self was both unknowable and not to be trusted, only actions that emanated from its two conflicting halves could successfully represent it in the world. That is to say, only divided action, action whose motivation is uncertain and attributable to at least two contrary causes, adequately reflects Lawrence's self. Such action destroys the heroic vision of integration through ironic fulfillment of its conditions. Yet it is a fulfillment, for it expresses truly the unknowable, untrustworthy, and divided self.

Conclusion

The outward orientation of Homeric man makes his identity essentially an exterior affair—of birth, status, heroic achievement, and recognition—rather than a subject for interior examination and revision. Odysseus largely defies this principle, relying least of Homer's heroes upon external determinants of his character and self-esteem. In his isolation he reflects deeply upon who and what he is. Eventually he becomes his own poet, recasting the truth of his experience in fictional form. But while Odysseus defies the rule of absolute external determination of self, he does not turn it on its head. He learns not to equate his identity with appearances, but neither does he divorce himself from the external world. He demonstrates a thoughtfulness more characteristic of "poetic" than "heroic" endeavor, but he does not divide himself into thinking and acting parts of radically different value. Odysseus can act in the world and believe that his actions express self. To the degree that action does proceed from thought—as when he leaves Ogygia to reenter the human world, or when he strings the bow to kill the suitors—Odysseus acts heroically.

The distinction between internal and external reality nascent in Homer's hero became in Lawrence the most crippling obstacle to his heroic confidence. Odysseus never has to examine the self as existing wholly apart from its actions; but Lawrence, who could not trust action, had no other choice. If the spirit is wholly independent of the flesh, as he sometimes thought, then the self can be neither known nor expressed in actions of the flesh. But if the spirit is associated with and corrupted by the flesh, as seemed the case to him at other times, then it must be fundamentally evil, and no action in the world can be pure. Either way, action becomes subject to profound distrust. Unless the self is known to be pure, the individual assertion inherent in heroic action will fail the basic heroic requirement of integrity. The heroic act becomes a source of guilt, a development inconceivable in Homer's world. Instead of feeling pride in the fine work he achieved with the Arabs, "I was continually and bitterly ashamed" (*SPW* 25). And instead of transforming his war career into a public life of fine achievements, he followed through the implications of his thought and degraded himself to a purely physical level by joining the ranks. He had learned that he could not succeed in both thought and action. His attempt to do so unloosed him from his spiritual moorings and drove him to abandon thought as the cause of his woes.

Lawrence's status as hero or villain sparks vigorous dispute even today, sixty-five years after his death. Whether Odysseus is truly heroic has exercised the critical imagination since the time of Socrates. According to the heroic model first established by the *Odyssey* and later embraced (if also modified) by Malraux and others, Odysseus ranks as the greatest hero of ancient times, greater than Achilles. He was a man of action who reflected on the source of action, its duties, and its consequences. He was also a poet, able to shape himself to the public so that it might recognize him as a human being and hero. Lawrence's case is more difficult: he acted significantly in extraordinary circumstances, and he reflected on action and self, recreating both in his "poetic autobiography."[73] But he also failed to achieve unity of self and act: the act was always in spite of the self, rather than stemming from it. Self-doubt destroyed belief for Lawrence, so that he could fully endorse neither himself nor the cause for whose advancement he acted. His postwar withdrawal, however, as an act proceeding from and expressing his loss of integrity, was in a sense a coherent act of self and as such an act of heroism. It may seem ironic that Lawrence should be recognized as a hero more for his retreat from public life than for his active participation in it. Yet his rejection of individual value—a rejection not merely verbal, but enacted through his acceptance of the life of a private airman—was an act of conviction and bravery. It was an act of profound cynicism as well, a cynicism that had penetrated his deepest being. Such, perhaps, is the fate of heroic action in the modern era.

Notes

1. 4.222–26; pp. 50–51. All translations from the *Odyssey* are Lawrence's unless otherwise indicated and are taken from T. E. Shaw, trans., *The Odyssey of Homer* (London: Oxford University Press, 1960). Book and line numbers refer to Homer's original, page numbers to Lawrence's translation.
2. See the letter to E. Garnett, August 26, 1922, in David Garnett, ed., *The Letters of T. E. Lawrence* (New York: Doubleday, Doran & Co., 1938), 360. This volume of letters will be referred to hereafter as Garnett.
3. By "poetry" I intend not the putting of words into verse but the fuller sense of a "making" or "creating," after the ancient term *poiesis.* I thus intend the word to refer broadly to the activity of aesthetic reflection in general, and to be synonymous with "art."
4. In book 4 we learn that Menelaus will have to face his, however, by living eternally with Helen in the Elysian Fields: it is not a happy prospect if the dreary life in Sparta is any indication of what lies ahead. Perhaps such thoughts are what finally drove Odysseus away from Calypso's divine island.
5. 12.184, my translation.
6. See Pietro Pucci, *Odysseus Polytropos: Intertextual Readings in the Odyssey and Iliad* (Ithaca: Cornell University Press, 1987), 60, and John Peradotto, *Man in the Middle Voice: Name and Narration in the Odyssey* (Princeton: Princeton University Press, 1990), 51–52.
7. *The Best of the Achaeans: Concepts of the Hero in Archaic Greek Poetry* (Baltimore: Johns Hopkins, 1979), 237, quoting W. J. Verdenius, "AINOS," *Mnemosyne* 15 (1962), 389.
8. Ibid., 240.
9. See C. Segal, "*Kleos* and Its Ironies in the *Odyssey,*" *L'Antiquité Classique* 52 (1983), 38–43.
10. And in both cases he lands on an island (Thrinacia in the inner story, Ithaca in the outer) where young men imprudently slaughter and eat cattle belonging to another more powerful entity, and meet their fates in doing so.
11. J. Notopoulos, "The Tragic and the Epic in T. E. Lawrence," *Yale Review* (1965), 338.
12. Ibid., 337.
13. Ibid., 337.
14. Ibid., 337.
15. Ibid., 331.

16. Besides Sophocles, Dante, Tennyson, Joyce, and others have recreated the Odysseus character in substantive aesthetic works. See W. B. Stanford, *The Ulysses Theme* (Ann Arbor: University of Michigan Press, 1968).
17. For the many transformations of Lawrence's legend, see S. Tabachnick and C. Matheson, *Images of Lawrence* (London: Jonathan Cape, 1988), and W. Chace, "T. E. Lawrence: The Uses of Heroism," in J. Meyers, ed., *T. E. Lawrence: Soldier, Writer, Legend* (New York: St. Martin's, 1989), 128–60. On "polytropic," see the next note.
18. Notopoulos, 338.
19. *Polutropos* means "of many ways," "of many turns," "of many devices," "resourceful," etc. The word is used in the first line of the *Odyssey* to identify Odysseus, whose actual name is not given until line 21. The adjective occurs one other time in the poem, at 10.330 and is unique to Odysseus.
20. Letter to B. Rogers, January 31, 1931, in Garnett, 710.
21. T. E. Lawrence, *Seven Pillars of Wisdom: A Triumph* (Garden City, NY: Doubleday, 1935), 549. Citations to *Seven Pillars of Wisdom* will hereafter be indicated in the text as *SPW.*
22. T. E. Lawrence, *The Mint* (New York: Norton, 1963), 195. The Norton edition has the same pagination as the original Doubleday edition. Citations to *The Mint* will hereafter be indicated in the text as *M.*
23. Letter to G. B. Shaw, August 17, 1922, in Garnett, 356.
24. Letter of February 20, 1924, in Garnett, 455–56. Also see his letter to E. Garnett, December 1, 1927, in Garnett, 550, and his accounts of the historical aspect of the book in Robert Graves and B. H. Liddell Hart, *T.E. Lawrence to His Biographers Robert Graves and Liddell Hart* (Garden City, NY: Doubleday, 1963), 79, 117–18 of the Graves section. Even if we accept at face value his claims (not maintained with perfect consistency even by himself) to historical accuracy, we would have to doubt them in the face of the enormous research that calls parts of his narrative into question.
25. Letter to L. Curtis, December 22, 1927, in Garnett, 559; quoted in M. D. Allen, *The Medievalism of Lawrence of Arabia* (University Park: Pennsylvania State University Press, 1991), 137–38.
26. "Evolution of a Revolt," in T. E. Lawrence, A. W. Lawrence, ed., *Oriental Assembly* (London, 1939), 131.
27. Letter of August 28, 1928, in Garnett, 622.
28. Letter to E. Garnett, September 7, 1922, in Garnett, 366.
29. The first notable incident of this kind occurs when he assumes the role of executioner. *Seven Pillars of Wisdom,* 181–82.
30. André Malraux, "Lawrence and the Demon of the Absolute," *Hudson Review* 8 (1956), 520.
31. It is difficult to know what Lawrence told Thomas, what Thomas heard from others, and what he might have invented. In any event, Thomas's account in *Friends* rings true: "In the course of these consultations I frequently asked him whether certain anecdotes I had gathered were true. . . . He

would laugh with glee and reply: 'History isn't made up of truths anyhow, so why worry?'" (184–85; and see 177–86 generally).

32. In a letter to Charlotte Shaw he wrote: "For fear of being hurt, or rather to earn five minutes respite from a pain which drove me mad, I gave away the only possession we are born into the world with—our bodily integrity. . . . You may call this morbid: but think of the offence, and the intensity of my brooding over it for these years." Letter of March 26, 1924, in Malcolm Brown, ed., *T. E. Lawrence: The Selected Letters* (New York: Paragon House, 1992), 261–62.
33. In another version of the incident Lawrence writes of "the breaking of the spirit by that frenzied nerve-shattering pain which had degraded me to beast level when it made me grovel to it, and which had journeyed with me since, a fascination and terror and morbid desire, lascivious and vicious, perhaps, but like the striving of a moth towards its flame." Oxford text, p. 285, quoted in Phillip Knightley and Colin Simpson, *The Secret Lives of Lawrence of Arabia* (New York: McGraw-Hill, 1970), 246.
34. See the letter to W. F. Stirling, June 28, 1919, in Brown, 165–66; and his letter home about a month after the alleged event in M. R. Lawrence, ed., *The Home Letters of T. E. Lawrence and His Brothers* (Oxford, 1954), 344.
35. John Mack observes that Lawrence "rarely told stories about himself that had no kernel of truth." "T. E. Lawrence and the Uses of Psychology in the Biography of Historical Figures," in L. Carl Brown and Norman Itzkowitz, eds., *Psychological Dimensions of Near Eastern Studies* (Princeton: Darwin Press, 1977), 52.
36. Malraux, 525.
37. Ibid., 525.
38. Ibid., 525.
39. Ibid., 529.
40. Letter of October 23, 1922, in Garnett, 371.
41. Robert Graves and B. H. Liddell Hart, eds., *T. E. Lawrence to His Biographers Robert Graves and Liddell Hart,* 172.
42. Irving Howe, "T. E. Lawrence: The Problem of Heroism," *Hudson Review* 15 (Autumn 1962), 356.
43. Jeffrey Meyers, *The Wounded Spirit: A Study of the* Seven Pillars of Wisdom (New York: St. Martin's, 1973), 130.
44. Letter to W. Rothenstein, October 20, 1932, in Garnett, 749
45. See, for example, Garnett 550 and *Seven Pillars of Wisdom,* 549.
46. Thomas J. O'Donnell, *The Confessions of T. E. Lawrence: The Romantic Hero's Presentation of Self* (Athens: Ohio University Press, 1979), 13. See also B. H. Liddell Hart, *Colonel Lawrence: The Man Behind the Legend* (New York: Halcyon House, 1934), 352, and Erik Lönnroth, *Lawrence of Arabia: An Historical Appreciation,* trans. Ruth Lewis (London: Vallentine, Mitchell, 1956), 50.
47. Stephen E. Tabachnick, *T. E. Lawrence* (Boston: Twayne, 1978), 106.
48. Letter of December 28, 1925, in *Home Letters,* 360–61.

49. Robert Graves and B. H. Liddell Hart, eds., *T. E. Lawrence to His Biographers Robert Graves and Liddell Hart,* 74.
50. William Chace, "Uses of Heroism," 139.
51. See the "Translator's Note" that accompanies Lawrence's *Odyssey.*
52. Letter to R. Graves, February 4, 1935, in Brown, 520–22.
53. Letter to E. M. Forster, August 28, 1928, in Garnett, 625.
54. Letter of November 22, 1923, in *Home Letters,* 356.
55. Letter to E. M. Forster, October 22, 1932, in Brown, 468. It is interesting that Lawrence should choose this particular analogy, because, while income tax forms are indeed impersonal, filling them out involves revealing private information.
56. Letter to H. Williamson, May 3, 1930, in Brown, 440.
57. Letter to W. Rothenstein, October 20, 1932, in Garnett, 749.
58. See also Allen, *Medievalism,* 164–65, and O'Donnell, *Confessions,* 4.
59. Howe, 358.
60. A letter written shortly before his death suggests that Lawrence recognized all too well his own presence behind the translation's façade: "I have my grave doubts about that book," he writes. "The translation is too unfaithful: too deliberately unfaithful." (Letter to R. White, April 13, 1935, in Brown, 532.) Typically, once he realized the influence of his personality on the work, he recoiled again from it but in a different direction. Now it became his fault, not Homer's, if his version read badly. At one point he expressed pleasure at hearing of a negative review, which he asked a friend to forward him: "About time somebody stood up for Homer," he wrote. "I've wanted to do it myself." (Letter to Liddell Hart, June 14, 1934, in Garnett, 810.)
61. Ronald Storrs, *The Memoirs of Sir Ronald Storrs* (New York: Arno Press, 1972), 472. Storrs saw Lawrence's "heightenings" and "colour" not as faithful attempts to render Homer, but as "most un-Homeric puckishness of his own" (ibid.). Storrs's only mistake lies in calling such puckishness "un-Homeric."
62. George Steiner, "From Caxton to *Omeros:* The Continuing Appeal of Homer to Anglo-Saxon Ideals and Experience," *Times Literary Supplement,* August 27, 1993, 15.
63. H. A. Mason, *To Homer Through Pope: An Introduction to Homer's* Iliad *and Pope's Translation* (New York: Barnes & Noble, 1972), 127.
64. See his letter to Forster of August 6, 1928, in Garnett, 618–22, in which he discusses *The Mint* and compares it to *Seven Pillars of Wisdom.*
65. On Lawrence's use of "we" in *The Mint,* see K. Hull, "Lawrence of *The Mint,* Ross of the R.A.F.," *South Atlantic Quarterly* 74 (1975), 340–48. O'Donnell also discusses the Lawrencian "we" in *Confessions,* 142, 172–75.
66. Storrs, *Memoirs,* 475.
67. Perhaps it is important to remind ourselves in this regard that although Lawrence's father had married, he had not married Lawrence's mother. The allusion to the Greek epitaph thus acquires further resonance.
68. Letter to Forster, August 6, 1928, in Garnett, 620.

69. The opening of the *Odyssey* hints at a similar attempt in Odysseus' case, insofar as he "ached in an agony to redeem himself." But Lawrence soon found that he could not carry through such a conception of Homer's hero. In the case of the RAF he was more persistent.
70. *M,* 217. The chapter beginning with this paragraph is especially thick with the religion of the air; see *M,* 217–19.
71. Robert Graves and B. H. Liddell Hart, eds., *T. E. Lawrence to His Biographers Robert Graves and Liddell Hart,* 75.
72. R. P. Blackmur, "The Everlasting Effort: A Citation of T. E. Lawrence," in *The Expense of Greatness* (New York: Arrow Editions, 1940), 20.
73. Tabachnick, 29.

Chapter Five

Lawrence's *Odyssey*

A "Prosaic" Approach to Greatness

Stephanie Nelson and Maren Cohn

> —You are a delusion, said roundly John Eglinton to Stephen. You have brought us all this way to show us a French triangle. Do you believe your own theory?
> —No, Stephen said promptly.
> —Are you going to write it? Mr. Best asked.
>
> —James Joyce, *Ulysses*

T. E. Lawrence's translation of the *Odyssey* is a very funny one. But then that is appropriate, since the *Odyssey* is a very funny poem. This is true in both senses of the word: the *Odyssey* is funny in that it is highly idiosyncratic, in particular in contrast to the *Iliad,* and it is funny in that, properly read, it should make one laugh.[1] The two senses are not unrelated—either in Homer or in Lawrence.

According to Longinus, the *Iliad* is sublime, the *Odyssey* merely a *muthos,* a tale, or a romance—albeit the romance of a Homer (*On the Sublime,* 9.11–15). Lurking behind the ancient critic's formulation is the idea that the older poem largely embodies the tragic mode, while the *Odyssey,* as he says, resembles a comedy. This is true in more respects than the "comedy of manners" to which Longinus points. The *Iliad,* after all, ends with Hector's funeral and the promise of Achilles' death, while the *Odyssey* ends, if not in a marriage, at any rate in a marriage renewed. Perhaps even more significantly, the *Iliad* aims at a tone that, however rich and deeply shaded, is, like the deep blue of twilight, a single piece. The *Odyssey,* in contrast, veers almost drunkenly between a world peopled by giants, witches, and mermaids, and one where an old nurse, a pig-herd, and a bunch of unwanted guests are the main actors. The poem, like its hero, turns in many ways, ranging from a seer's terrifying vision of the great hall drenched in blood to the near slapstick of Athena's ornithological transformations, to the profoundly moving and profoundly mature undertone of Penelope's need for Odysseus and of Odysseus' for Penelope. It is in this sense, or in all of these senses, that the poem is most truly a comedy.

A Prose Tradition

Lawrence's *Odyssey* uniquely captures the fullness of Homer's comic tone. The translation is, in equal and overlapping parts, brilliant and bizarre. Within it Lawrence moves from the most elevated and precious "Wardour-Street" diction to the most colloquial and even vulgar language:[2] from "Wherefore the Sun-God blotted out / The day of their return" (proem; 1.7–8) to "Mentor, you crazy mischief-maker . . ." (21; 2.243);[3] from "Forth from the lovely waters sprang the sun into its firmament of brass . . ." (28; 3.1) to "Telemachus, here is no room for false modesty: no room at all." (28; 3.14). Lawrence can even make the switch in successive phrases: "Tell us therefore of yourselves, O guests! Is yours a business venture, or do you cruise at random . . . ?" (30; 3.71–72) or "But Ino of the slim ankles had seen him,—Ino the bright, a daughter of Cadmus. She had been born mortal in the beginning: just a simple-speaking girl . . ." (78; 5.333–35) or "Then he drew himself up and with a billet of wood spared from his chopping he clubbed the beast. It died" (205; 14.425–26).[4] The mortar for this strange construction was necessarily prose. The mortar, however, also located Lawrence within a tradition that already implied certain assumptions about the work to be translated.

Translations of Homer fall into one of two camps: prose or poetry, and this truth about Homer applies even more fully to the *Odyssey* in particular. The *Odyssey* is great poetry, but it is also, in some ways unlike the *Iliad,* and as the multitude of children's versions of the *Odyssey* testify, a rollicking good tale. In our own time, in the prose school of Rouse, Rieu, and Shewring, focus has fallen upon the *Odyssey* as a story. On the other hand, however, beginning (perhaps) with Lattimore, and continuing on through Fitzgerald, Fagles, and, at the extreme, Christopher Logue, translators have attempted to capture the poetry of Homer. This division between prose and poetry would have confronted Lawrence in an even more pronounced form. Poetic translations of Homer, the tradition that Lawrence rejected, stretched back to the great voices of Chapman, Dryden, and Pope. The rejection placed Lawrence squarely in the other camp, a tradition that went back through Samuel Butler's *Odyssey for English Readers* to William Morris's translation to Butcher and Lang. These works, often intended for schoolboys, cared mostly about the plot. Hiller's "idiomatic" American version (1927) grounds the prose approach most succinctly, if not elegantly: "the modern youth with his present-day vocabulary, being more concerned with a story itself than with the appropriateness of its diction, finds archaic expressions dry and uninteresting, and hence the thrilling tales of the *Odyssey* . . . are now not generally read by the young . . ."[5] But behind all this lurk Butcher and Lang.

Butcher and Lang's translation, the schoolboy staple of Lawrence's day, twists its prose into a shape that neither God nor mortal had ever spoken. Their chosen idiom presupposes a King James who had established in his Bible not a genius of cross-traditions, but a twelve-step program for translators. Butcher and Lang are nothing if not consistent, but what they finally adhere to is a wholly artificial language created by themselves. Ironically, modern Homeric scholarship might seem to validate exactly this artificiality. Since the work of Milman Parry and Albert Lord, it has become increasingly clear that the language of the Homeric epics was never a "natural language" spoken by a people, but rather a poetic language developed over generations of oral minstrelsy for particular stories in a particular meter.

Although unaware of what has now become the standard understanding of the "oral tradition" from which Homeric poetry sprang, Butcher and Lang sensed in Homer a deeply traditional language that they likened to the language of the King James Bible.[6] In one respect they were right: Homer's artificial language also founded an entire culture and a great literary tradition. In another respect, however, they were not. The language of Homer may not have been "natural," in the sense that it was not a native spoken language; but its artificiality, unlike Butcher and Lang's, did not entail monotone confinement within a single, unvarying form. Placed in what is to us a very peculiar position, given a language that could not expand naturally as its native speakers developed it, the poets of the Homeric tradition found their own way to develop their language, not through the change of progressive growth but through self-reference. The language of the *Odyssey* is the language of the *Iliad,* just as the language of Butcher and Lang is in a way the language of the King James Bible. The difference is that Butcher and Lang take their model not only seriously, but solemnly as well. The *Odyssey* does not.

Lawrence's very particular style and, as we argue below, his very particular use of Butcher and Lang create a translation that, despite Lawrence himself, enables him to bridge the two camps of prose and poetry. Poetic translations traditionally have focused on the language of Homer, prose translations on the story. Lawrence was quite emphatic about locating himself in the latter camp. He was no poet, he said, and he did not feel that Homer was the place to begin.[7] The point of the *Odyssey* was the tale: it was "the oldest book worth reading for its story and the first novel of Europe." "Obviously the tale was the thing . . ."[8] And yet, in conveying the story as vividly, concretely, directly, and perversely as he could, Lawrence managed also to capture linguistic qualities in the *Odyssey* that have disappeared under the hands of far more accomplished poets and far more consistent translators than Lawrence either aimed, or was able, to be.

The Humor of Discrepancy

The comedy of the *Odyssey* may escape us, in part because of the distance of language and time, and in part because of the solemnity of scholars. If we read attentively, however, it will shine out nonetheless. Samuel Butler—despite Lawrence's dislike of his translation, which he thought "too little dignified"[9]—was a noteworthy iconoclast in his own right and one of the first modern readers of the *Odyssey* to point out the poem's distinctly irreverent undertone. Consider, for example, Butler's discussion of Penelope:

> Sending pretty little messages to her admirers was not exactly the way to get rid of them. Did she ever try snubbing? Nothing of the kind is placed on record. Did she ever say, "Well, Antinous, whoever else I may marry, you may make your mind easy that it will not be you." Then there was boring—did she ever try that? Did she ever read them any of her grandfather's letters? Did she sing them her own songs or play them music of her own composition? . . . Did she ask Antinous or Eurymachus to sit to her for her web—give them a good stiff pose, make them stick to it, and talk to them all the time? Did she find errands for them to run, and then scold them and say she did not want them? or make them do commissions for her and forget to pay them, or keep on sending them back to the shop to change things, and they had given ever too much money and she wished she had gone and done it herself? . . . With one touch of common sense the whole fabric crumbles into dust.[10]

The comic note has been picked up by some more modern translations. Thus Fitzgerald describes Athena, disguised as a young girl and leading Odysseus to Phaeacia, as "the awesome one in pigtails" (7.44), and has her speak, with malice aforethought, to the middle-aged Odysseus as "good grandfer, sir" (7.31).[11] The phrases translate nothing in the Greek. They do nonetheless capture precisely the mood of the scene, a mood underlined when Odysseus, who seems here oblivious to Athena and her machinations, proves to have known all along who the child guide really was (13.323).

Homer sometimes uses similes to achieve his effects. Odysseus, at the final moment of his adventures, swept from the shipwreck that follows the eating of Helios's cattle, clutching for dear life at a branch as he hangs above the whirlpool Charybdis, waiting for the bits of timber that make up his life raft to reappear, is compared to a judge—and we learn that jokes about lawyers have a longer history than we might have thought:

> To me, longing for them, they came
> Late—just like when a man stands up out of the lawcourt, for dinner,
> having judged all the quarrels of the hot-headed youth,
> so long it took the timbers to appear again out of Charybdis. (12.438–41)[12]

Or again, Odysseus, having arrived home, but not truly home, lies in agony as a stranger in his own house, listening to his maidservants laughing as they go out to sleep with the men usurping his place, torn between his need to assert his identity and the impossibility of acting. And in this epic anguish Homer likens him, tossing and turning, to sausages grilled over a fire (20.25–28).

Lawrence does not make Homer funny by merely translating the funny bits for laughs. In fact, Lawrence can be quite solemn in such places. His Odysseus remarks on his late rescue from Charybdis, for example, in a rather dignified tone: "Very late it was before my hopes were answered and they came: as late as the supper-rising of a justice from the courts, who has stayed to settle the many suits of his hot-blooded litigants" (180–81). Rather, whether deliberately or not, Lawrence has taken on more generally the attitude of the *Odyssey.* He views not just the funny bits but all of the poem with a sardonic eye. The view, like the refreshing cynicism of Butler, does not point out that there are parts of the *Odyssey* that are quite comic. Instead, its overall vision of the poem enables the reader to see that there are.

Lawrence's humor differs from Butler's in his use of language. Butler's is a completely simple, even colloquial, language. His Athena brings up the subject of Odysseus, for example, thus: "Then Minerva said, 'Father, son of Saturn, King of kings, it served Aegisthus right, and so it would anyone else who does as he did; but Aegisthus is neither here nor there; it is for Ulysses that my heart bleeds, when I think of his sufferings in that lonely sea-girt island, far away, poor man, from all his friends'" (2; 1.44–50).[13] The impetus for such a translation becomes clear when we compare Butcher and Lang: "And the goddess, grey-eyed Athena answered him, saying: 'O father, our father Cronides, throned in the highest; that man assuredly lies in a death that is his due; so perish likewise all who work such deeds! But my heart is rent for wise Odysseus, the hapless one, who far from his friends this long while suffereth affliction in a sea-girt isle, where is the navel of the sea, a woodland isle, and therein a goddess hath her habitation . . . '" (2).[14]

As Guy Davenport puts it, when in doubt "Butcher throws in a 'yea' . . . and Lang throws in a 'lo.'"[15] The surprise is Lawrence: "Swiftly there took him up Athena, goddess of the limpid eyes. 'Our Father, heir of Cronos, Lord of lords! That man Aegisthus has been justly served. May everyone who slaughters a victim after his fashion go down likewise into hell! But my heart is heavy for Odysseus, so shrewd, so ill-fated, pining in his misery of exile on an island which is just a speck in the belly of the sea'" (2). In places Lawrence's translation seems almost a mere negotiation between Butler and Butcher and Lang. Thus, for Lawrence, it neither "served Aegisthus right" nor did Aegisthus "assuredly [lie] in a death that is his due"—rather, he "has been justly served"; while again, Athena's heart neither "bleeds" nor "is rent" for Odysseus, but simply "is heavy." But such obedient restraint is not always the case. On the ornate side Lawrence's "goddess of the limpid eyes" is more elaborate even than Butcher and Lang, while for colloquialism, Butler's "lonely sea-girt island" and Butcher and Lang's "sea-girt isle, where is the navel of the sea, a woodland isle" becomes for Lawrence, quite wonderfully, "just a speck in the belly of the sea."

Similarly, when Telemachus spots the disguised Athena, Butler has "Telemachus saw her long before anyone else did. He was sitting moodily among the suitors thinking about his brave father, and how he would send them flying out of the house, if he were to come to his own again and be honored as in the days gone by. . . . 'Welcome,' said he, 'to our house, and when you have partaken of food you shall tell us what you have come for'" (4; 1.113–24). Butcher and Lang are more elegant: "And godlike Telemachus was far the first to descry her, for he was sitting with a heavy heart among the wooers dreaming on his good father, if haply he might come somewhence, and make a scattering of the wooers there throughout the palace, and himself get honour and bear rule among his own possessions. . . . 'Hail stranger, with us thou shalt be kindly entreated, and thereafter, when thou hast tasted meat, thou shalt tell us that whereof thou hast need'" (5).

But for Lawrence, "[Telemachus] sat despondent in the hurly, fancying to himself his honest father's sudden arrival from somewhere, somehow: and the scatter there would be, through the palace, of these wasters when they saw him stride in to regain men's respect and king it honorably once more over his household. As so he dreamed amidst the unheeding suitors he became aware of Athena waiting by the threshold. . . . 'Accept, O guest, the friendliest greetings. Enter and taste our food: and thereafter make known to us your every need'" (4). Here Lawrence's version is a little of both. His Telemachus is "dreaming," not "sitting with heavy heart," but he is doing so "amidst the unheeding suitors," not just "among the suitors." He greets his guest not just with a "Welcome" but with an "O guest," and if he does not declare "thereafter, when thou hast tasted meat, thou shalt tell us that whereof thou hast need" neither does he want to know simply "what he has come for." But perhaps most powerful in this passage is the way that Lawrence has used his Butcher and Lang backdrop to set off not only the elements that need emphasis—"Enter and taste our food," for example, or "despondent in the hurly"—but also the tone of Telemachus' own daydreams, in which his father "strides" and "kings it honorably," coming from—a sudden shift and a crucial one—"somewhere, somehow."

The wonder of Lawrence's language lies in its ability to be not simply neither Butler nor Butcher and Lang, but to be neither by being both of them. The trick benefits him immediately, enabling him to use his Butcher and Lang background to highlight, to enormous effect, his Butlerisms. Thus, as Athena describes Odysseus to Zeus, she says in Butler: "This daughter of Atlas has got hold of poor unhappy Ulysses, and keeps trying by every kind of blandishment to make him forget his home, so that he is tired of life, and thinks of nothing but how he might once more see the smoke of his own chimneys" (2; 1.55–59). In Butcher and Lang the same lines run: "[Atlas's] daughter it is that holds that hapless man in sorrow: and ever with soft and guileful tales she is wooing him to forgetfulness of Ithaca. But Odysseus yearning to see if it were but the smoke leap upwards from his own land, hath a desire to die" (3). Lawrence, once more keeping a Butcher and Lang background, makes a serious innovation. To begin slightly earlier in the passage: "This wave-beset wooded island is the domain of a God-begotten creature, the daughter of baleful Atlas whose are the pillars that prop the lofty sky: whose too are the deepest soundings of the sea. The daughter has trapped the luckless wretch and with subtle insistence cozens him to forget his Ithaca. Forget! Odysseus is so sick with longing to see if it were but the smoke of his home spring up, that he prays for death" (2). The Butcher and Lang—the "wave-beset," the "domain," the "baleful," the "soundings," and the "lofty sky"—sets off the wonderful moments in which Lawrence hits home—the "sick with longing," the "prays for death," and the great "Forget!"

In his mixing of styles and voices Lawrence in fact takes his cue from the "O-Homer," as he termed him, even if he also plays some variations on the given theme. The poet of the *Odyssey,* whether or not he composed the *Iliad* as well, had only that language at hand. As we will see below, it is a language he used in deeply subversive ways. Lawrence was in the opposite position. He, and he chafed at the bit at it, was trapped into the translator's role, into the episodes, the events, and the similes of his poet.[16] What he was not trapped into, as Homer in many ways was, was a particular style of language. His response was a stroke of brilliance: he neither simply established his own idiom, as Butcher and Lang had done, nor reverted to a purely colloquial tone, as Butler had. To adopt either course would have been, finally, simply to replace one formulaic language with another. Rather, Lawrence set up a very posh Butcher and Lang style, but set it up as a norm allowing him the opportunity to subvert it. It is here that the spirit of Lawrence's *Odyssey* appears. It is not quite where the spirit of Homer's *Odyssey* appears—but then, perhaps, that is the point.

In some scenes, the odd variations in Lawrence's language do in fact reflect a similar undertone in Homer. Consider, for example, Lawrence's highly variable treatment of Penelope. When the heroine first appears in the poem, beckoned from her room by the songs of the poet Phemius, Homer describes her, as Lawrence has it: "So, like a stately goddess among mortals, she descended upon the suitors: to halt there where the first great pillar propped up the massy roof. As veil for her face she held up a fold of her soft wimple: and the ever-watchful maidens covered her, one on either side. Thus stood she and wept . . ." (10; 1.328–35). The "stately goddess" contrasted with the utterly simple "Thus stood she and wept" captures perfectly the Homeric combination of authority and helplessness that Telemachus is about to challenge. Similarly, when Penelope later hears of Telemachus' departure she declares: "O cruel women whose hearts knew all, but did not think to call me back from my bed when the lad went down to his black hollow ship. If I had known that he was intending the journey, very surely he should have stayed, however eager: or gone only by leaving my dead body behind him in the halls" (65; 4.729–34). The "my dead body" here, as set off by the overall Butcher and Lang tone of the "cruel women whose hearts knew all" and the "black, hollow ship," may ring too colloquially for more conventional ears—and yet its vehemence goes a long way towards explaining why it was that Telemachus insisted so strongly, and unconvincingly, that his mother not be told of his voyage.

One of the nicest touches comes in book 18 as Penelope decides to elicit further gifts from the suitors. Her sad tale begins, in Lawrence's translation: "Listen, when he left . . ." (253; 18.259). The phrase itself is pure Lawrence, but the pseudo-conspiratorial tone is clearly in Homer as well. Odysseus, unknown and reduced to beggary in his own home, hears with his own ears his wife claim she soon must marry again. His response is delight (18.281). Odysseus knows a good story when he hears one.

These examples, however felicitous, are nonetheless the least of Lawrence's contribution to Homer. The best of it lies not in his respect for the text, but precisely in his disrespect.[17] Lawrence's irreverence follows that of the *Odyssey,* but this hardly makes him a slave (as he was quite determined not to be) to his author.[18] In fact he often, if not always, subverts Homer in just the places the *Odyssey* does not subvert the *Iliad.* This subversion emerges particularly in the somewhat cynical view he takes of the *Odyssey,* a view he pointedly expresses in the Translator's Note, where he complains that "only the central family stands out, consistently and pitilessly drawn—the sly cattish wife, that cold-blooded egotist Odysseus, and the priggish son who yet met his master-prig in Menelaus" (vii). Lawrence's own cold-bloodedness in making such assessments is still capable of shocking. Bernard Knox, who introduces the latest edition of the translation comments apologetically: "It looks as if the long, drawn-out labor on the translation, much of it accomplished after a full day's work on his service assignments, had ended by filling him with a distaste for the whole enterprise. . . . Happily, these feelings do not surface in the translation."[19] Whatever its cause or its justification, however, Lawrence's sardonic view has an unanticipated benefit: it desanctifies the poem, and by doing so shows us its characters stripped of any reverence accumulated by age. In the *Peloponnesian Wars* Thucydides complains (without citing Herodotus by name) of the tendency of history, through time, to "fight its way into myth" (21.10). So had the *Odyssey* done for most modern readers, until such men as Butler and Lawrence looked at it with a pitiless eye.

In this respect Lawrence's vision of the *Odyssey* serves the poem precisely by virtue of its irreverence, by stripping away the dull patina of dust and grime that solemnizes the epic actors and reduces their bright colors to a somber palate of heroic browns and grays. If episodes from the *Odyssey* all of a sudden appear funny in Lawrence's translation, it is often not because he has written them to be so, but because by subverting all our expectations of what a decent sober translation should do, he has suddenly confronted us with them as if it were for the first time. Suddenly, through the oddity of Lawrence's medium, the actors of the *Odyssey* appear; they cease to have a consistent heroic tone; they cease to be heroes and become instead human beings.

Here too Lawrence, whether deliberately or not, does not betray Homer. The great heroes of the *Iliad* appear, in the cold, clear air of the *Odyssey,* in a different light. Here Achilles desires not the glory of immortal fame, but the rejected choice, an inglorious, and even the most inglorious life (11.465 ff.).[20] Nestor too has been transformed. From the wise statesman of the *Iliad* he has become a grossly garrulous old man, one who answers Telemachus' question, "Have you heard of Odysseus's fate?" not by the appropriate answer "No," however embellished, but by a discourse that begins at sunrise (3.1–3) and ends: "So he spoke, and the sun went down and the darkness came over them . . ." (3.329). It is no surprise that Telemachus, on his return, should appeal to his newfound friend, Nestor's son Peisistratus, not to make him go back and visit the old man, for fear he never get away again (15.195–214).

Despite Lawrence, the Odysseus of the *Odyssey* is not merely a cold-blooded egotist with a sly cattish wife and a priggish son—though there are elements of truth in these judgments. But the Helen and Menelaus of the *Odyssey,* gaily trading tales of the good old days in Troy until Helen finds it expedient to drug the drinks; and Agamemnon, with his obsession about his wrongs and his injured status; and Ajax, with whom Odysseus might have reconciled had he not found more interesting things to do; and, overall, the suitors' quite simple dismissal of the possibility of the great hero's return (e.g., 2.246–51) are as debunking as anything in Lawrence. For Homer, as for Lawrence, the great epic world of heroism was one thing, and real life, as lived in Ithaca here and now, was quite another.[21] Lawrence derives a great deal of humor from the discrepancy between the two. And so does Homer.

A Place for Disrespect

One way to see that the *Odyssey* is basically a comedy is to look at how it primarily achieves its humor—through incongruity and a juxtaposition of opposites. The method is very much the mainstay of comedy as we know it, in Hamlet's encounter with the grave-digger, in the comedy of Byron's rhymes ("intellectual," for example, with "hen pecked you all,") or in the Wizard of Oz's admonition to "pay no attention to the man behind the curtain." We have seen this method already in Homer's occasionally disconcerting similes and Lawrence's occasionally pointed irreverence. The same form of comedy, one that results from sudden shifts in vision, runs throughout both Homer's poem and Lawrence's translation.

Sheer and incongruous physicality, for instance, a comic element fully familiar to us, pervades the *Odyssey.* It has often been remarked, not least amusingly by Butler, that Odysseus spends the majority of his time concerned with food, drink, sex, and sleep—none of which (perhaps excepting the third) is usually seen as particularly a heroic activity. Hence when Sancho Panza feels the need to convince Don Quixote that he is not in fact enchanted, he points out that the Don still feels the need to (as Sancho puts it) "make the big and little waters" (Part 1, ch. 48–49). For Achilles, in a very different mode, the dark night of the soul that he undergoes over the death of Patroclus implies necessarily that he will not eat—despite Odysseus' urging (*Il.*19.205–31). When Vergil, a rather more solemn author than his predecessor, took over the story of a visit to Hades, Aeneas' companion is cast overboard by divine forces demanding the sacrifice of an innocent for the safekeeping of the rest of the expedition. The aspect of the story that Vergil failed to include is the grosser one; the Odyssean counterpart to Palinurus, Elpenor, lands in Hades because he was too drunk, or hung over, to notice that the best way to get down from the roof was by the ladder.

This purely physical humor, the humor of *Don Quixote,* of the Three Stooges and of Aristophanes, recurs throughout the *Odyssey.* Again and again, Homer makes a point of the physical. The physical humiliation of the beggar Irus, like the humiliation of Thersites in the *Iliad,* although repellant to our tastes, is presented with relish. A similar delight appears in the suitors' graphic descriptions of what Irus and/or Odysseus will suffer at the hands of the boogie monster, King "Hold-um" (18.85; 21.308). Odysseus repeatedly volunteers that the demands of the belly, like the demands of life itself, will not allow us the leisure of our own sorrow (e.g., 7.215–221; 18.52–54). The *Odyssey,* like the *Iliad* and like the *Aeneid,* sees that there is a discrepancy between the gross needs of the body and the world of heroism. For the *Odyssey* what this means is that the great world of heroism is simply going to have to wait.

Lawrence, in this respect, differs strikingly from the poet of the *Odyssey.* Where the *Odyssey* is (*mutatis mutandis*) nearly as happy in the realm of the physical as Aristophanes, Lawrence's deep distaste for the body shows through the crevices of his translation. Thus in Lawrence's translation, Odysseus's account of the belly's needs ends up somewhat stilted:

> But instead I will ask leave to obey my instincts and fall upon this supper, as I would do despite my burden of woe. See now, there is nothing so exigent as a man's ravening belly, which will not let him alone to feel even so sore a grief as this in my heart; but prefers to overwhelm his misery with its needs for meat and drink, forcibly and shamelessly compelling him to put its replenishment before his soul's agony. (100; 7.215–21)

It is the "soul's agony," not the "exigency" of the belly, that Lawrence prefers to focus on.

Lawrence, however, does not thereby miss Homer's sense of the discrepancy between the grandly heroic and the down to earth. Rather it emerges in Lawrence not so much in a sense of the physical as in an insistence on the concrete and specific, even when the insistence belies the Greek.[22] Aegisthus, in seducing Clytemnestra, "was ever speaking in the ear of Agamemnon's wife, trying to steal her love" (36)—a far more concrete rendering than the literal "he often charmed the wife of Agamemnon with his words" (3.264). Odysseus fears, in approaching Nausicaa, "that he might outrage her modesty," not that "the maiden might become angry in her heart" (88; 6.147). Euryalus "sneered in his face," not "quarreled face to face" (108; 8.158), to which Odysseus replies not "O Stranger" but "you, whoever you are" (108; 8.166). "Speaking in a small voice he spoke a word to me" becomes "he hissed at me sharply" (207; 14.492), while "I lay in it [the cloak] gladly" becomes "I snuggled in" (207; 14.502). And finally, in the boar hunt,

Lawrence's picture of Odysseus "pressing hard upon the pack, his poised spear trembling in his eager hand" creates a scene very unlike Lattimore's "went close behind the hounds, shaking his spear far-shadowing" (266; 19.438).

The insistence on concreteness makes sense. After all, Lawrence, unlike most translators of Homer, had lived a markedly Homeric life, and his early work on the translation found him living in remote regions of the Middle East. That the need to find a vivid physical vision of each scene should happen to accompany a present or remembered setting in which such a vision was readily available was mostly by chance. The chance, nonetheless, worked well with the original. When Lawrence describes the approach to the city of the Phaeacians, he calls it a "fairway"—and so it seems to have been: "I saw that beach in Phaeacia, with the building ships on the right, and the drawn-up ships on the shingle to the left, with the crowds moving down the ridge between them. That was why I said fairway, because the comers had to thread the lines of ships."[23]

Most importantly, however, Lawrence's sense of the *Odyssey*'s physical grounding seems to parallel Homer's use of the grosser physicality of food, drink, and sleep to undercut the epic grandeur that should have been Odysseus' due. Nowhere in the *Iliad* does Achilles eat simply because he is hungry, nor fall asleep simply because he is tired. Odysseus' fate, in contrast, is shaped by sleep and hunger—while carrying the winds of Aeolus, in the Cyclops's cave, on the Island of the Sun. Lawrence's specificity captures exactly this sense of the critical importance of the mundane. His rendering of Telemachus' reproach to the suitors: "Do our family affairs jar your sense of niceness?" (18; 2.138) grounds the moment far more clearly than the conventional "if your *thumos* takes offence at these affairs" (2.138). His Autolycus does not merely "excel all men in thieving and oath-taking" (19.395–96), but "swore falser and stole better than all the world beside" (265). And perhaps best of all, Odysseus' men, in Lawrence, do not merely grieve at being turned to swine, they "grieved, squealingly" (142; 10.241).

A similar form of humor in the *Odyssey* stems from a sort of comedy that Aristophanes also revels in, the kind that Mephistopheles introduces into Goethe's *Faust:* that of puncturing pretensions. A great, if minor, character in this regard is Telemachus' newfound friend Peisistratus. As Helen and Menelaus, accompanied by Telemachus, grow profound over the loss of old friends, Peisistratus points out that he too has lost a brother—although one he didn't know—but nonetheless, he prefers to eat his dinner without tears (4.190–200). As Telemachus, reminded by Athena of the need to return home, springs to his feet ready for instant action, Peisistratus reminds him that it is only two A.M. and the bus doesn't leave until eight—or at least the epic equivalent (15.48–55). Similarly, Telemachus, imprisoned in Ithaca, besieged by the suitors and incapable of asserting his newfound manhood, finally finds refuge in divine aid. Athena, miraculously, comes to the rescue: she borrows a boat (2.386–87).

Lawrence has, if anything, an even better eye for priggishness than Homer. His Menelaus does not merely describe his riches to Telemachus, he catalogues them: "See, out of the store of treasures ranged in my house I give you the fairest and costliest:—Item, a wrought mixing bowl of solid silver doubled with gold about the rim. Work of Hephaestus. Hero Phaedimus, King of Sidon, endowed me with it . . ." (62; 4.613–18). Menelaus knows that one mustn't challenge the gods, but he doesn't mind claiming his own among mortals: "Dear children, with Zeus no mortal can vie . . . On earth—well, there may be a man as rich as myself or there may not . . ." (46; 4.78–81). His welcome of Telemachus—"Wonderful, wonderful"—suggests the hearty but distracted exclamation, the too-vigorous handshake—and the attendant quick loss of interest (49; 4.169). And his welcome of Peisistratus is worthy of even Polonius: "As the son of such a father naturally you speak with wisdom. It takes no art to pick out the offspring of a man into the texture of whose days the son of Kronos has woven bliss in the marriage bed and in the procreation of fair children. Nestor, to return to him, has had supremely granted to him for all his days . . ." (50; 4.206–209).

Lawrence's inclination to puncture pretension, whether it be in Menelaus, in (at least as he sees it) Homer himself, or in the classical tradition, perhaps stems from an inclination to distrust authority. The same inclination appears in Butler, whose lively reading of the *Odyssey* seems to relate directly to his rebellion against all that is Victorian. In Lawrence, the feeling may have a very different, and perhaps a much more complex, grounding. The effects, however, very much resemble each other. And the result is almost entirely positive—although the "almost" here is one that will finally need to be watched.

Homer himself shows little enough reverence for what in his world represents the greatest authority—the gods. In fact the gods are perhaps the single most dependable source of humor in the *Odyssey.* With the caveat that there is, finally, a profound sense in Homer of the fateful working of the universe, and so of the divine—a sense that Lawrence may or may not share—as a translator Lawrence has ventured into a Homeric territory where most translators (not to mention scholars) have feared to tread.

On some level we are certainly meant to understand Athena's appearance to Telemachus as singling him out for a particular destiny. There is, however, another, quite comic, aspect to their relation. Athena bolsters Telemachus' confidence but also assures him that in general sons are never as good as their fathers (2.276–77). This hardly seems designed to reassure. Later, as she leaves Nestor and Telemachus in the form of Mentor, she explains, in detail, the need to set off early to collect a debt "neither recent, nor small" (3.367–68)—and flies away in the form of a sea-hawk.[24] Noemon is mystified by the puzzle of how Mentor could both be with Telemachus in Pylos and at home on the beach (4.653–56). And when Athena, again in the form of Mentor, finally does come miraculously to assist Odysseus against the suitors, her aid consists in upbraiding the hero for his newfound cowardice—and turning him into a starling. Given such a goddess, Telemachus' rather skeptical reply to his father's assertion, that with helpers like Zeus and Athena two men could battle 118 suitors, seems rather more plausible than not. As Lawrence puts it, "Telemachus said gently: 'This pair rule all mankind, with the Immortals to boot, and are indeed mighty helpers; but their seat is very far away among the clouds'" (227; 16.262–65).

There are no consequences for Homer's gods, as there are for men; for them everything, finally, is a comedy. No better commentary on this fact can be found than one Homer himself offers through Demodocus' song about Aphrodite's infidelity with Ares, and Hephaestus' revenge. The song is deeply applicable to Odysseus, who has lived through the results of Helen's unfaithfulness and must now be wondering about his own wife's relation to her young and wealthy suitors. And yet his response is to laugh. What among men provides a source of tragedy is, among the gods, only a source of laughter (8.326–69).

Lawrence, in his own rebellion against authority, finds the perfect comic touch for the gods, although perhaps for different reasons. His Hermes complains to Calypso about the hardships of his journey in exactly the petulant tones of Homer's god: "by no choice of my own did I fare to you across so unspeakable a waste of salt water. Who would willingly come where there is no near city of men to offer sacrifice to the gods and burn us tasty hundreds of oxen?" (71–72; 5.100–102). His Zeus objects to Athena's dramatic lament: "My child, too fierce are the judgements of your mouth. Besides, I think this last move was of your scheming . . ." (69; 5.22–23). And in Lawrence, the great resolution of the *Odyssey,* which undoes precisely the unending cycle of revenge dramatized in the *Oresteia,* finds, as Zeus waves his magic wand, a suitable fairy-tale tone: "Let the parties compose a binding treaty, by virtue of which [Odysseus] shall remain their king: while we will expunge from memory this slaughter of the people's kith and kin. So shall they love one another as of yore and peace abound, with wealth" (326; 24.283–86). "And Pallas, still with Mentor's form and voice, set a pact between them for ever and ever" (327; 24.546–48).

Fighting One's Way Out of Myth

We have seen enough to warrant the claim that the humor of the *Odyssey,* like the humor of Lawrence's translation, is a humor of discrepancy. One of the most obvious ways in which this appears has been explicitly noted by translators as far back as Butler: "The *Odyssey* (as every one knows) abounds in passages borrowed from the *Iliad*" (viii). As Lawrence puts it: "His generation so rudely admired the *Iliad* that even to misquote it was a virtue" (Trans. Note, vii).

To Butcher, Lang, and Butler, the *Odyssey*'s use of lines from the *Iliad* could not but smack of deliberate allusion at best and plagiarism at worst. From our modern understanding of the oral tradition, the matter appears quite differently—and yet there are places where Robert Graves's accusation that the poet of the *Odyssey* tends to quote the *Iliad* "very naughtily"[25] rings true. As Telemachus makes his first venture into public life, for example, he, like Achilles in the opening of the *Iliad,* hurls down the scepter that represents the great lineal descent of kingship—and bursts into tears (2.80–81). The scene in which Nausicaa, too shy to mention her hopes for marriage before her father, claims she must look after her brothers by doing their laundry is amusing in itself. For those who recognize in the line with which she takes out the wash the far more solemn line with which the body of Hector was carried out for burial, the humor takes on a darker shade. As Telemachus addresses Penelope on her first appearance in the *Odyssey* he reproaches her for paying too much attention to poets, and declares: "Speech will be the care of the men / All the men, but me most of all; for mine is the power in this house" (1.358–59). The line, ironic in itself, becomes far more so when we recognize it as Hector's final parting from Andromache: "War will be the care of the men / All the men, but me most of all, who live in Ilium" (6.492–93). And perhaps out of mere naughtiness, but perhaps with a more telling reference: as Odysseus, home for the first time in twenty years, washes off the grime of battle, the line used for how the water was heated is the line that describes how Achilles washed, with warm water, the corpse of Patroclus. There is the absolute finality of loss, and the recovery from loss, in the *Odyssey* as in the *Iliad*—but here, as always, it occurs on far more mundane a level.

The "naughty" self-referentiality in Homer's language is the deepest level on which the humor of juxtaposition works. This sense of incongruity (which, for Kant at any rate, was the essence of humor) pervades the poem, and in its many varied aspects makes up its essence, giving it an essentially comic nature. In the end we come to see that the various forms of humor we have catalogued are grounded in what may be the single most basic element of the *Odyssey,* the *Iliad.* Just as a little girl in pigtails is funny not in herself, but because she is Athena, and as sausages roasting on a fire are funny not inherently, but because they are intended to depict a great epic hero in his moment of agony, the *Odyssey* is funny and at the same time profound, precisely because it is not the *Iliad.* The *Odyssey*'s use of its predecessor's lines does not demean or ridicule the pathos of the *Iliad*—but it does point to a truth that is, within its comic vision, even deeper. The *Iliad* presents, in many ways, a vision of tragedy, the soul's bleak journey in which man, stripped of all accessories, must alone confront the blank face of his own mortality. The *Odyssey* is about something else: the fact that after the black night of the soul is over, we all, as we know only too well, still get out of bed the next morning and go to work.

In this way the humor of Lawrence's translation, as it brings out the discrepancy between the heroic world and the world as we know it, reflects the deepest level of the poem. The humor stems largely from the juxtaposition of the heroic and the mundane, and draws not only on the *Iliad* but also sets up pregnant possibilities within its own sphere: thus we find the parallel, for example, between Telemachus' and Odysseus' journeys—one, as it were, a journey into space, to distant unknown worlds, the other a trip to Hoboken. On the comic level the poem splits one epic world into two and presents Odysseus' task as that of fighting his way out of the heroic but frozen world of the *Iliad,* the world of Nestor and Menelaus and Helen, of the shades of Achilles and Agamemnon and Ajax, through a magic world of giants, goddesses and witches, and into the perhaps comic, perhaps vulgar, but nonetheless living world of Ithaca, his wife, and his son.[26]

Redemption: The Final Parting of Poetry and Prose

For T. E. Lawrence, or rather, for "Lawrence of Arabia," a poem about the possibility of fighting one's way out of the land of myth back into ordinary life must register. The positive impact on his translation of this resonance, and with it his profound distrust of his own, or anyone's, heroism, create the skepticism and irreverence that so deeply captures the spirit, even when it almost deliberately avoids the letter, of Homer. Lawrence's connection to the themes of the poem also has, however, a negative impact, the response that emerges in his obviously polemic but finally deep-rooted vision of Odysseus as "that cold-blooded egotist" and Penelope as his "sly, cattish wife."

In the closing stage of the *Odyssey,* Lawrence and Homer seem finally to part ways. Lawrence frankly regarded the latter half of the poem as a bore: "Perhaps the tedious delay of the climax through ten books may be a poor bard's means of prolonging his host's hospitality," he wrote (Trans. Note, vii). Even more harshly, he remarks on books 23 and 24: "His last movement drools on and on like one of Schubert's, everybody (author included) dying to end it, but mellifluously unable."[27] His growing impatience with the second half of the *Odyssey* is reflected in the translation as well, most particularly in an ever-increasing tendency to condense Homer's lines and formulae into fewer and fewer words. Thus in book 1 Lawrence translates: "A maid came with a precious golden ewer and poured water for them above its silver basis, rinsing their hands. She drew to the side a gleaming table and on it the matronly house-keeper arranged her store of bread and many prepared dishes, making an easy grace of all the hospitality" (4; 1.136–40); while in book 17 exactly the same lines are translated: "the maid of the ewer poured for them the hand-water over its basin and drew up their table which the house-keeper hospitably spread with loaves and many dishes." (234; 17.91–95). Again, in book 21 Lawrence condenses "for your lady mother did not give birth to you as the sort of man who would become a drawer of the bow and arrow" into "In very truth your lady mother, when she conceived you, was not making a master-bowman" (285; 21.172–73).[28]

Lawrence had been at work on the translation for a long time, nearly twice as long as he had anticipated. He had good reason quite simply to be growing tired of the job. There is, however, another element in play here, one reflected in the continuing cynicism of lines like "As Odysseus had run on, furnishing her with proof too solid for rejection . . ." (309; 23.206), or Penelope's coldly official request, "Bed is yours the instant your heart wills, for have not the Gods restored you to your own great house and native land? But now that Heaven has put it in your mind, tell me of this ordeal remaining. Later I must know, and forewarned is forearmed" (311; 23.257–62). In such passages Lawrence's style maintains its rather detached and sardonic view of its protagonists—but in so doing, it now deviates from Homer.

The cynical, debunking side of the *Odyssey* may well extend into the reunion of Odysseus and Telemachus and the ambivalent scene in which Telemachus just fails, at Odysseus' behest, to master his father's bow. Odysseus' conquest of the suitors is also far from unambiguous, as Athena's comic ornithological transformations on the one hand, and, on the other, Homer's vivid description of a feast defiled by blood point out. But whatever else is sardonic in the *Odyssey,* the reunion of Odysseus and Penelope is not. What may be the one moment in the poem in which its hero speaks unguardedly and without premeditation is also the one moment in the poem in which its poet cannot be accused of speaking with two voices.

What matters in their recognition scene is that Penelope acts upon Odysseus, not the other way around. Penelope tricks Odysseus not into revealing the factual secret known only to the two of them—no trick, after all, would have been necessary for this—but into revealing, despite himself, how much the secret matters. It is not Odysseus' knowledge of how their bed was made, but rather his deep need that the bed, and the marriage it was built for, remain fixed, that provides the emotional climax of the *Odyssey.*[29] Odysseus, for the first time, speaks unguardedly, and so does Homer.

But Lawrence does not. With one great exception—"Who has changed my bed?"—Lawrence's version of the scene shows us a Penelope who speaks "to draw her husband out" and an Odysseus who, in response, is "ruffled," all against the background of a still thoroughly Butcher-and-Langian prose:

"This she said to draw her husband out; and indeed Odysseus was ruffled into protesting to his wife: 'Woman, this order pains my heart. Who has changed my bed? It would task the cunningest man—forebye no God happened to shift it in whim—for not the stoutest wight alive could heave it up directly. That bed's design held a marvelous feature of my own contriving'" (309; 23.183–89). Similarly, in the recognition itself, Odysseus "runs on" and "furnishes proof" in a context of pompous formality: "'That was the style of it woman, as I explain: but of course I do not know if the bed stands as it did; or has someone sawn through the olive and altered it?' As Odysseus had run on, furnishing her with proof too solid for rejection, her knees trembled . . ." (309; 23.206–207).

The secret that has been shared all along by Odysseus and Penelope is an important revelation to the reader: the stable center of Odysseus' world was not the storeroom where Penelope weeps over her husband's bow, but the bed that Odysseus fixed so that it could never be moved. It was to this that Odysseus has been returning, and to this that he will return again. It was for this that he fought his way out of the world of heroic myth into the world of banal reality.[30] It is also this, finally, that Lawrence refuses to acknowledge. There is ultimately a redemption in the *Odyssey* that does not appear in Lawrence's translation. And the redemption is deeply rooted in the *Odyssey*'s cynical view of heroism, a view that moves, finally, from simple debunking to a profound sense that it is better to be a live dog than a dead lion. Odysseus may still retain to a degree the immortality and glory of Achillean heroism, but his more powerful and lasting place in human memory is that of a purely human dependence on other humans. Lawrence did not translate this latter part. It may be that he did not translate it because he did not know it.

The reasons why T. E. Lawrence might be drawn to the *Odyssey* are many and have been amply recognized. The *Odyssey,* very much unlike the *Iliad,* presents a hero whose anti-heroic qualities seriously rival his heroic status—a resonant situation for one who himself had such trouble accepting and living out a heroic model. The *Odyssey,* too, concerns a hero who has an almost pathological problem with being named; again the relevance to Lawrence, or Ross, or Shaw, is not far to seek. Odysseus is a hero and so a leader of men, and yet his men, as has often been pointed out, exist primarily to be killed in the course of his adventures. Odysseus, very much unlike Achilles, is a hero who lies, who pretends to be what he is not, and who is, in Lawrence's words, a "cold-blooded egotist" capable of sacrificing the interests of those who believe in and depend upon him to his own inscrutable ends. And so, to his own infinite agony, was Lawrence.[31]

Not only does Lawrence's experience deeply resemble Odysseus', in certain respects it also resembles Homer's. As we have seen above, the poet of the *Odyssey* was as much a follower and a debunker as Lawrence. But Odysseus and the "O-Homer" were also, emphatically, creators, and this is a role Lawrence consciously chooses to reject. In *Seven Pillars of Wisdom* he writes:

> I followed and did not institute; indeed, had no desire even to follow. It was only weakness which delayed me from mind-suicide, some slow task to choke at length this furnace in my brain. I had developed ideas of other men, and helped them, but had never created a thing of my own, since I could not approve creation. When other men created, I would serve and patch to make it as good as it might be; for, if it were a sin to create, it must be a sin and shame added to have created one-eyed or halt.[32]

In the end, Lawrence refuses to be a poet in the fully creative, redemptive sense of the word. In his Translator's Note he attributes to Homer a diffidence and smallness that he finds in himself ("this Homer lived too long after the heroic age to feel assured and large," he writes; "his work smells of the literary coterie"), but these are qualities that most readers have trouble actually locating in *Odyssey.* Given Lawrence's resistance to the generative powers of poetry, his choice of prose for the translation acquires greater resonance and serves to distinguish him from both Homer and his hero. For the latter figure, at the climactic moment of regaining his identity, is described as a poet. In Lawrence's version:

> Odysseus the master of craft had by now handled and surveyed the great bow up and down. Calmly he stretched it out with the effortless ease of a skilled musician who makes fast both ends of a piece of twined cat-gut and strains it to a new peg in his lyre. Changing the bow to his right hand he proved the string, which sang to his pluck, sharp like a swallow's cry. (291; 21.404–11)

It is not a coincidental comparison. Throughout the Odyssey the hero repeatedly re-creates his own story, and with it his own identity: he can shape himself as Nobody to the Cyclops; to the Phaeacians as Odysseus, known for all craft and whose fame reaches the heavens; and as "Contention, son of Uncaredfor, son of All-Woe" to his father (24.305–306). He is a man whose crisis of identity stems exactly from his own power with words. Lawrence, too, shaped himself through the use of words, and it would be fair to say that his personal crisis also stems from his verbal power—it was the writing of *Seven Pillars of Wisdom,* after all, that sent him into his mental tailspin after the war. But when Odysseus is compared to a poet, thus becoming fully the poet or "maker" of his own identity, he parts company with Lawrence. Lawrence refuses to accompany him on that poetic journey. His medium is prose.

Lawrence's insistence on prose, as simple and reasonable as it was, thus implies a distance both from the hero of the poem and the poem's author. It is a distance underlined by Lawrence's own comments. Despite their similarities, Lawrence refused in the end to identify himself with the hero of his poem, or that hero's redemption. If Homer, by reworking the epic world of the *Iliad,* was in some ways as much of a translator as Lawrence, he also, by committing himself to his own new creation, finally became something else. Like Cervantes, who began a parody of romantic chivalric novels and created Don Quixote, "O-Homer" discovered new worlds almost despite himself. Lawrence, in the end, did not.

Given the echoes of the *Odyssey* in his own life, it would be surprising if the chance assignment of translating the poem had not awakened in Lawrence some deep responses. It would also be surprising if such responses did not bring out undertows and submerged elements in the poem, which had (and have) failed to create ripples in the translations of more conventional scholars. It is those ripples that we have been concerned with here. Lawrence of Arabia was, in ways that other essays in this volume have considered, a uniquely appropriate translator of the *Odyssey,* as the *Odyssey* was a uniquely appropriate poem for his translation. One wonders, however, whether the results of the fit were what his publisher had sought when looking for a translator who, granted, would do something different from older versions, "a free translation—rather a new interpretation of the *Odyssey.*"[33] In Lawrence's hands what emerges in Homer's *Odyssey* is not a true sense of heroism but a deep suspicion of heroism, and not the flavor of foreign places and great deeds but discrepancy, incongruity, and doubt. In many ways, in fact, and as his Translator's Note warns us, Lawrence's *Odyssey* ends up more as an anti-*Odyssey.* Which may also be why it both is, and is not, so appropriate a translation of the poem.

Notes

1. On the question of humor in the *Odyssey,* in particular in relation to Lawrence, see H. Hazlitt, "On Translating Homer," review of Lawrence's *Odyssey,* in *The Nation* 135 (December 21, 1932); Robert Graves, "Colonel Lawrence's *Odyssey,*" in *Steps: Stories, Talks, Essays, Poems, Studies in History* (London: Cassell, 1958); Maurice Bowra, "Two Translations," review of *The Odyssey of Homer,* by T. E. Shaw (Lawrence), in *The New Statesman and Nation,* April 8, 1933, 449; H. A. Mason, *To Homer Through Pope: An Introduction to Homer's Iliad and Pope's Translation* (New York: Barnes & Noble, 1972), 10; Maurice Hewlett, trans., *The Iliad of Homer* (London: The Cresset Press, 1928), vi, ix. Lawrence had read and approved of Hewlett's notes (Letter to B. Rogers, April 16, 1928, in D. Garnett, ed., *The Letters of T. E.*

Lawrence [New York: Doubleday, Doran & Co., 1938], 589). Lawrence himself, despite the humor of his translation, never finally accepted that Homer meant to be funny.

2. As the Penguin edition of *Seven Pillars* has E. M. Forster saying on its back cover, Lawrence's is "a style nicely woven of Oxfordisms and Doughty" (T. E. Lawrence, *Seven Pillars of Wisdom,* Penguin Books, 1962).
3. *The Odyssey of Homer,* T. E. Lawrence, trans. (New York: Oxford University Press, 1991), 21. When citing Lawrence's version we will give both his page numbers and the conventional book and line numbers.
4. Other examples: "give us a lead," "make a match of it," "tittle," "made wise," "high and dry," "where did she pick him up?" "Be mum now"; and on the other hand: "and lo!" "verily," "laved," "yea," "nay," "guerdon," "pursuivant," "sore perplexed," "bedesman," "was like to have been," "Reiver," "squire," "knight," "assoil," "privily," "sacrings," "hauberk," "welkin," "beldam," "caitiff," "forbye," "wight," and the list goes on.
5. *The Odyssey of Homer,* trans. Robert H. Hiller, Winston Companion Classics (Philadelphia: John C. Winston Co., 1927), vi.
6. Hence Hiller, vi: "The poems of Homer are written in a peculiar poetic dialect which bears to the later classical Greek somewhat the same relation as does the language of the King James version of our Bible to modern English."
7. "I have never written a line of poetry in my life . . . and I don't feel that Homer is the thing to begin on." Letter to B. Rogers, April 16, 1928 in Garnett, 598.
8. "Translator's Note" to Lawrence's *Odyssey,* v, vii.
9. Letter to R. Isham, January 2, 1928, in M. Brown, ed., *T. E. Lawrence: The Selected Letters* (New York: Paragon House, 1992), 362. Lawrence also remarked that Butler failed to convey "all the picturesque side: the bric-à-brac: and most of the poetry has evaporated with Homer's queer, archaic, dignity." Letter to C. Shaw, June 4, 1928, quoted in J. Wilson, *Lawrence of Arabia: The Authorized Biography of T. E. Lawrence* (New York: Atheneum, 1990), 829.
10. Samuel Butler, *The Authoress of the Odyssey,* with a new introduction by David Grene (Chicago: University of Chicago Press, 1967), 130.
11. *The Odyssey,* trans. Robert Fitzgerald (New York: Farrar, Straus and Giroux, 1998). Similarly, in the introduction to Fitzgerald, see Donald Carne-Ross on Helen: "a domestic figure but no doubt dressed in full rig and accompanied by two maid-servants bringing her golden distaff and a silver basket holding her yarn. . . . She is still as full of herself as she was in the *Iliad,* and freely admits, not without satisfaction, that her conduct has left much to be desired. Oh I was terribly wicked, I know, but how bravely they all fought for me!" (xiv)
12. Unless otherwise noted, translations are our own.
13. *The Odyssey:* Rendered into English prose for the use of those who cannot read the original, Samuel Butler trans. (London: A.C. Fifield, no date).
14. S. H. Butcher and A. Lang, *The Odyssey of Homer Done into English Prose* (New York: Macmillan, 1888).

15. "Words to Butcher and Lang are invariably decorative, swatches of color all. So that dogs *tramp*. And when there is a paucity of adjectives for fringe, Butcher throws in a 'yea' ('many were the men whose towns he saw and whose minds he learnt, yea') and Lang throws in a 'lo' ('Lo, the dogs withhold him from his way')." "Another Odyssey," in *The Geography of the Imagination* (Nonpareil, 1978), 29–44.
16. Lawrence regretted having simply, as he put it on one occasion, to "expose the fraud" of Homer's excessively artful, "creeping," and "booky" work. (Letter to W. Rothenstein, October 20, 1932, in Garnett, 749.) At the same time, however, his attitude toward his rôle as a translator was complex and doesn't admit of simple characterization, for he was often grateful to be a mere follower without responsibility for creation. See note 18.
17. See Bowra, "Two Translations," 449: "we are forced to admit that [Lawrence's] judgements [of Odysseus and Penelope in the Translator's Note] are not undeserved," and further: "if the Odyssey is to be read without cant and self-deception, keen insight of this kind is invaluable."
18. Here again the complexity of Lawrence's attitude towards translation emerges. On the one hand he sought out translation as a way of avoiding artistic independence: "In translating you get all the craftsman's fuss of playing with words, without the artist's responsibility of their design & meaning." Letter to E. M. Forster, August 28, 1928, in Garnett, 625. "No original stuff, of course: just translations. I hope not again to do anything of my own. It is not good for a man to make things." Letter of November 22, 1923, in M. R. Lawrence, ed., *The Home Letters of T. E. Lawrence and His Brothers* (New York: Macmillan, 1954), 356. On the other hand, the Translator's Note in itself witnesses his refusal simply to cloak himself under Homer's mantle, and at times he acknowledges that he couldn't help adding touches of his own to the translation: "I'm always trying to underline, to score heavier, to put in little bits to try and lift the thing to life." Letter to B. Rogers, January 29, 1930, quoted in J. Wilson, "T. E. Lawrence and the Translating of the *Odyssey*," *Journal of the T. E. Lawrence Society* 3 (Fall 1994), 55. Lawrence may well have submitted to the task of translating as humbly as he could, but perfect humility was not in the end possible for a nature such as his.
19. B. Knox, Introduction to Lawrence's *Odyssey*, xvi. Perhaps it should be added that Knox's last opinion is open to some question.
20. Douglas J. Stewart, *The Disguised Guest: Rank, Role, and Identity in the* Odyssey (Lewisburg: Bucknell University Press, 1976), 60–63, compares the position of Achilles in Hades to Falstaff's attitude towards honor: "The point in both cases being that death makes a hero, a man honored for his dying bravery, and thus in a sense there are no heroes; there only have been heroes!" (61). William G. Thalmann, *The* Odyssey: *An Epic of Return* (New York: Twayne Publishers, 1992), 120–21, points out that the poem's return to Hades in book 24 reinforces the connection of heroism and death, and in so doing contrasts this conception of heroism to Odysseus.

21. Cedric H. Whitman, "The *Odyssey* and Change," in *Twentieth Century Interpretations of The* Odyssey, ed. Howard W. Clarke (Englewood Cliffs, N.J.: Prentice-Hall, Inc., 1983), 90–99, sees Odysseus as representing the old world in the new. See also Anne Amory, "The Reunion of Odysseus and Penelope," in Charles H. Taylor, Jr., ed., *Essays on the* Odyssey: *Selected Modern Criticism* (Bloomington: Indiana University Press, 1969), 111; and David Grene, "The *Odyssey:* An Approach," in *Midway,* 1969, 47–68: "The smoke from the chimney, the bed made from the tree anchored in the room, the crooked paths up and down the hills of Ithaca—these are descriptions where the perceived externals fuse with some deeper aspect of meaning. They have nothing to do with the Sirens or the strait between Scylla and Charybdis" (57).
22. Lawrence wrote to E. M. Forster, August 28, 1928 about *The Mint* (in Garnettt, 624): "What you say about the emphasis I get on simple words. . . . [M]ayn't it be partly because I do try & feel every article or emotion which comes into the book? I tie myself into knots trying to re-enact everything, as I write it out. It's like writing in front of a looking-glass, and never looking at the paper, but always at the imaginary scene." The comment clearly applies to his translation work as well.
23. Letter to B. Rogers, January 26, 1935, in Garnett, 842.
24. See also Grene, "Approach," 61, on Athena's proclivity toward ornithological transformations.
25. Graves, "Colonel Lawrence's *Odyssey,*" 187.
26. See Pierre Vidal-Naquet, *The Black Hunter: Forms of Thought and Forms of Society in the Greek World,* trans. Andrew Szegedy-Maszak, (Baltimore: Johns Hopkins, 1986), 15–38.
27. Letter to B. Rogers, August 20, 1931, in Garnett, 732.
28. In the first seven books, Lawrence requires between 100 and 110 percent the number of lines as are in the Greek; in the next seven books, except for book 13 (on which he plainly lavishes fond attention), Lawrence requires between 90 and 100 percent of the lines to translate a given number of Homeric verses, while in the final ten books he requires between only 80 and 90 percent. The following chart illustrates the breakdown by book:

    Book 1:489 lines Lawrence to 444 Homer (110%)
    Book 2; 473 Lawrence to 434 Homer (109%)

Book 3: 551 to 497 (111%)	Book 14: 511 to 533 (96%)
Book 4: 897 to 847 (106%)	Book 15: 461 to 557 (83%)
Book 5: 525 to 493 (107%)	Book 16: 373 to 481 (78%)
Book 6: 349 to 331 (105%)	Book 17: 481 to 606 (79%)
Book 7: 354 to 347 (102%)	Book 18: 354 to 428 (83%)
Book 8: 570 to 586 (97%)	Book 19: 500 to 604 (83%)
Book 9: 555 to 566 (98%)	Book 20: 344 to 394 (87%)
Book 10: 542 to 574 (95%)	Book 21: 391 to 434 (90%)
Book 11: 608 to 640 (95%)	Book 22: 454 to 501 (91%)

Book 12: 443 to 453 (98%)	Book 23: 313 to 372 (84%)
Book 13: 496 to 440 (113%)	Book 24: 492 to 548 (90%)

29. See Kevin Crotty, *The Poetics of Supplication: Homer's* Iliad *and* Odyssey (Ithaca: Cornell University Press, 1994), 198–99. For the importance of Penelope's deception of Odysseus, and so the revelation of his vulnerability, see Anne Amory, "The Reunion of Odysseus and Penelope," 111; W. B. Stanford, "Personal Relationships," 26, in Taylor, *Essays on the* Odyssey; and Marilyn A. Katz, *Penelope's Renown: Meaning and Indeterminacy in the* Odyssey (Princeton: Princeton University Press, 1991), 165, 176–77.
30. See Katz, 179–80, for the centrality of the bed.
31. It may not be a coincidence that Lawrence can translate Penelope's emotion, although not Odysseus', with no Butcher and Lang at all: " . . . her knees trembled, and her heart. She burst into tears, she ran to him, she flung her arms about his neck and kissed his head and cried: 'My Odysseus, forgive me this time too, you who were of old more comprehending than any man of men . . . '" (310). "Sly and cattish" as she may be, Lawrence seems to have had less trouble, finally, with her character.
32. T. E. Lawrence, *Seven Pillars of Wisdom: a Triumph* (New York: Doubleday, 1935), 565.
33. Quoted in Wilson, *Lawrence of Arabia,* 814.

Chapter Six

T. E. Lawrence and the Shaws

Jonathan Hart

I

On July 22, 1922, T. E. Lawrence wrote to George Bernard Shaw asking him to read a long book he had printed but not published: *Seven Pillars of Wisdom.* Lawrence's ambivalent attitude towards his work—he realized its importance but was disgusted by it—comes out in the letter. Lawrence claims "I am not a writer, and successfully passed the age of 30 without having wanted to write anything" (4). But he portrays himself as feeling compelled to have written the book: "I was brought up as a professional historian, which means the worship of original documents. To my astonishment, after peace came I found I was myself the sole person who knew what had happened in Arabia during the war: and the only literate person in the Arab Army. So it became a professional duty to record what happened" (4).[1] To some extent, I think it is right to take these words at face value. He did feel, at times, as if his writing were more of a duty than an opportunity. In his correspondence with the Shaws, in his dealings with publishers, and in his choice to turn his back on a fellowship at All Soul's, Oxford, for a life as an enlisted soldier, Lawrence repeatedly chose a route that was not opportunistic, comfortable, easy, or profitable.

What also arises in this first letter to Shaw is Lawrence's forthrightness, modesty and self-effacing nature that verges on self-denegration. Lawrence is rare: he knows how difficult it is to write well: "the job became too much for me" (4). There is, especially in Lawrence's correspondence with Charlotte Shaw over the next four years, a certain *agon* with GBS, the great writer of an earlier generation, an *agon* that is never far from the surface, even if that contest is most affectionate in tone. In this first letter to Shaw, Lawrence wrote: "Your first book was not perfect, though it was a subject you had chosen for yourself, and you had an itch to write!" (4). Lawrence says that he himself has cut down to his own size and stitched together "second-hand words," "borrowed expressions and adjectives from everybody I have ever read," and has included "a lot of half-baked thinking" (4). While Lawrence is attacking his own writing, he also undermines his own military claims: "I'm not the least a proper soldier" (4). But paradoxically, in his correspondence with the Shaws in this crucial period Lawrence again and again tries to establish himself both as a proper writer and soldier.

It is hard to imagine what Shaw made of this letter, but to give him credit, he ignored its convolutions and responded. But before I proceed to Shaw, I wish to return to Lawrence's labyrinthine shifts in tone and syntax. He dwelt in doubt and in conditional clauses: "If you read my thing, it will show you that your prefaces have been written in vain, if I'm a fair sample of my generation" (4). In case Shaw were to think that this "thing" were amusing, it is not: "it's long-winded, and pretentious, and dull to the point where I can no longer bear to look at it myself. I chose that moment to have it printed!" (5). Lawrence obviously understood irony and had some flare for the theatrical. Still, the self-criticism is striking; this tact is hardly the stuff of self-promotion.

And just when the convolution seems to be reaching a threshold, Lawrence adds a few extra twists, leaving such a maze that even Shaw might have been perplexed. The letter bears reading whole, but here is a long paragraph that gives some indication of what Lawrence was building up to:

> You'll wonder why, if all this is true (and I think it is) I want any decent person (still more a person like yourself).* [*ambiguous: but I wanted to avoid expressing my liking for your work.] to read it. Well it's because it is history, and I'm shamed for ever if I am the sole chronicler of an event, and fail to chronicle it: and yet unless what I've written can be made better I'll burn it. My own disgust with it is so great that I no longer believe it worth trying to improve (or possible to improve). If you read it or part of it and came to the same conclusion, you would give me courage to strike the match: whereas now I distrust my own judgement, and it seems cruel to destroy a thing on which I have worked my hardest for three years. While if you said that parts were rubbish, and other parts not so bad, and parts of it possible (and distinguished those parts variously) then your standards might enable me to clear up mine, and give energy enough to tackle the job again. (If you say it is all possible then I will reluctantly get rid of your own books from my shelves.) [5; insertion on "ambiguous" is Lawrence's own postscript]

These shifts are dizzying. First, Lawrence returns to his reason for publishing the work: it is the only history of these events. He then attempts to bond with Shaw as a reader but also to try to guess his reaction and almost to coerce it. Would he really be disappointed if Shaw said that the book was all good? Shaw is supposed to be Lawrence's "judgement" but Lawrence is almost coercing that judgment even as he gives him apparent freedom in a series of conditionals and concessive clauses. And even though he had already met the Shaws by chance at their London flat earlier that year, Lawrence seems to be wrestling with their reputation. Despite his shifts, however, Lawrence is also forthright: the book disgusts him, but it would be cruel to destroy something "on which I have worked my hardest for three years." If Lawrence is ambiguous, as he says, about Shaw, that ambiguity and ambivalence extends to his view of himself and his labors. Lawrence is not hiding behind the more usual stance of the uncertain and unpublished writer: I just dashed this off and I don't really care what happens to it.

Poor Shaw. But he could handle himself. He might have felt that he was on the front lines and Lawrence of Arabia had him in some intricate trap. The end of the letter closes in ever smaller circles:

> All this is very unfair—or would be, if you knew me: but deleting that twenty minutes with Cockerell we are utter strangers, and likely to remain so, and therefore there is no pressure on you to answer this letter at all. I won't be in the least astonished (indeed I'll write another of the same sort to a man called Orage whom I have never met, but whose criticism I enjoy): and my opinion of you will go up. Yours with many apologies.
>
> T. E. Lawrence.
>
> Incidentally: I don't want people to know the book exists. So whether you reply or not, I hope you will not talk of it.

Shaw is treated with great intimacy and then is told he is a stranger and does not need to respond and that besides, Lawrence was going to contact a critic he had not met. Lawrence also says that Shaw will go up in his estimation if he does not respond, but why write the letter in the first place? If Lawrence does not want anyone to know about the existence of his book, why is he writing Shaw and thinking about writing Orage? Perhaps Lawrence wants to try his luck with some influential readers without the public knowing about the project. This letter would leave most people, except Bernard Shaw, breathless, exhausted, and speechless.

In his reply of August 25, 1922, Shaw drops formalities and really ignores the multitude of shifts in Lawrence's letter. He makes two points:

> There is not the smallest doubt (within human limits) that any publisher would jump at your book on Arabia; and there is no doubt at all that the book, having forced itself from you, will be published with whatever imperfections its mortal lot may involve, whether you like it or not.

> Second, as man to man. It will be pure waste of your time for me to read your book before you send it to press; but it would not waste mine, as I shall read it sooner or later. (6)

Attempting to stick to the point and have Lawrence focus on the business of bringing his book out, Shaw seems also to try not to become involved with the intricacies of Lawrence's emotional state. He builds him up and urges him to use his time wisely. Moreover, Shaw wanted to put Lawrence at ease. Near the end of his brief letter, he sets out a more informal relation (just as he did in his form of address—"My dear Lawrence (to drop ceremony)"—to begin the letter): "You need not stand on any ceremony with us. You are a privileged soul, and can deal with both of us as with old friends" (7). Shaw speaks both for himself and for his wife, Charlotte, who was in Ireland and had only noted the meeting with Lawrence in her diary in March but had not corresponded with him.

Not surprisingly, Lawrence, on August 27, 1922, continues his contrary ways. He mimics Shaw's "First" and "second" construction and underlines the words. Then he quibbles about Shaw's use of "privilege": "and as you say I'm privileged, I'll try it: though it's an unwholesome state to live in: and I don't think I won there on my own efforts" (7). His doubt about his massive manuscript precedes by a few sentences his skepticism over his status as a war hero. There is enough modesty and self-doubt to fill many vessels; and this doubling doubt persists.

But for all the dizzying and difficult shifts, Lawrence does show some admirable self-awareness and an ability to put together in suggestive ways war and peace, the poet and soldier, these aspects of his life and writing that pressed upon him. In this same letter he mentions that Shaw was to be one of six people to read the book and explains why:

> You see the war was, for us who were in it, an overwrought time, in which we lost our normal footing. I wrote this thing in the war atmosphere, and believe that it is stinking with it. Also there is a good deal of cruelty, and some excitement. All these things, in a beginner's hands, tend to force him over the edge, and I suspect there is much over-writing. You have the finest cure for flatulence, and I have great hopes that you will laugh at parts of what I meant to be solemn: and if I can get at you before you have forgotten which they are, then I'll have a chance to make it better. (8)

Except for the self-satire, which might be a little too self-conscious, Lawrence's attempt to make sense of the war through writing and revision is understandable. His self-critical stance here is laudable, for he did not blind himself with the glory of his reputation as a soldier or assume that writing was easy and a reward for his heroic actions. This passage is one of the key places in Lawrence's letters: he knows he is overwrought; he sees the cruelty in war; he suspects that the experience of the war has, in places, adversely affected his writing. Even his ironic and satirical laughter at his own pretensions, as awkward or maddening as it could sometimes be, suggests a person who is interested in becoming pompous or in believing the myth that he had become. This humility is a quest for the marriage of his experience as a soldier and as a writer, sometimes achieved through the disciplining power of irony and satire directed toward himself. It is also something that he attempted to live as well as accomplish in his writing, revisions, and discussions of his experience as a soldier and as a writer. Lawrence did not accept the honors offered to him; instead, he sought to be an enlisted man. This became, in all its complications, an attempt to be a plain soldier, an ordinary man. In his writing Lawrence did not seek the riches Shaw said he could have with *Seven Pillars of Wisdom.* This behavior, as puzzling as it was to Shaw and other contemporaries and perhaps to some among us now, might have been owing in part to being overwrought and in part to being on a kind of spiritual quest for atonement. It is not easy to turn one's back on honors, position, wealth, and fame as Lawrence did. He could never quite escape them, however, and his correspondence with the Shaws shows his attempts to follow the daily routine of a soldier with one modest goal: to get his book out into the world in a form that pleased him.

In a letter to Bernard Shaw of September 30, 1922, Lawrence tries to sound breezy about his leaving the Colonial Office and enlisting in the RAF: "It's a cleaner living than Eastern politics, and I haven't a busman's wages of my own, so must do something" (10). Lawrence's aversion to eastern politics is understandable, but that Winston Churchill dismissed him penniless is highly doubtful (10).[2] Lawrence seems rather to have chosen his lot: his ongoing quest to be an ordinary man in the armed forces.

This is where Charlotte Shaw enters the scene once again. Shaw had given Charlotte *Seven Pillars of Wisdom* to read. In a letter to Sydney Cockerell of November 24, 1922, Charlotte Shaw talks about "a wonderful MS that has been given to us to read by someone you brought to see us" (10). Although Charlotte says she was sworn to secrecy, she gives enough of a hint that it would not be hard for Cockerell to guess Lawrence's identity. This is just one indication of how hard it was for Lawrence to hide behind obscurity and why Bernard Shaw thought that this was a perverse game, precisely because it was impossible.

In a letter of November 30, 1922, Lawrence says he was afraid that GBS was delaying because he did not wish to tell him that *Seven Pillars of Wisdom* was "rubbish," and that if Shaw wished to think it "rot" then Lawrence would "agree with you and cackle with pleasure at finding my judgement doubled" (11). Shaw replies, ignoring once more Lawrence's irony, and explains why he has not had a chance to read all of Lawrence's great tome: "My wife seized it first, and ploughed through from Alpha to Omega. It took months and months and months; but it carried her through" (11). A wit with his own store of irony, Shaw acknowledges Lawrence's genius but admits his own bafflement before it. He also hints at the apparent high-mindedness and impracticality of Lawrence's own attitude toward *Seven Pillars of Wisdom* and its appearance in print: "However, I know enough about it now to feel rather puzzled as to what is to be done with it. That, you may say, is your business; but the Life Force will take it out of your hands. Obviously there are things in it that you cannot publish. Yet many of them are things that WONT die. It would cost you too much to engrave the book on plates of gold, and bury it somewhere for somebody to dig up and start in business as a prophet" (11). Shaw suggests sealing the book so the Trustees of the British Museum can open it in a hundred years and supports his case with examples from Charles Dickens and Oscar Wilde, suggesting thereby that he considers Lawrence's work to be worthy of the best literary company. GBS leaves little doubt about how important he thinks *Seven Pillars of Wisdom* is: "It is one of the Cheop's pyramids of literature and history" (12). Clearly, Shaw understands the significance of Lawrence as a writer and historian. GBS also sees Lawrence as a great man: "You are evidently a very dangerous man: most men who are any good are: there is no power for good that is not also a power for evil" (12). While suggesting that Lawrence also provide an abridgment for "general circulation," he compares him to Caesar (12). In Lawrence's reply, on December 7, 1922, he states his admiration for Caesar's *Commentaries* but contrasts his own confessional style to its restraint: "They are the antithesis of mine: indeed I suspect that no successful general ever spilled so much of himself on to paper as I did" (15). Lawrence then mentions his enlisting in the ranks of the RAF, which he continues to maintain he did for money and that, should the abridgment be successful, he will abandon the service: "If so I'll become a civilian again. You have no idea how repulsive a barrack is as permanent home. It reconciles me to the meanness of the abridgement" (16). In fact, Lawrence often praises the service as a means to earn a basic wage and stay out of public view, a kind of monastic reflection on, or even penitence for, the war and the trying aftermath of working for the terms of peace. His enlistment as an ordinary serviceman is a contentious point for which Lawrence, even as here in the same letter, often shifts motives.

GBS continues discussing with Lawrence the publication of *Seven Pillars of Wisdom.* Shaw tries hard to place the book with his own publishers, Constables, telling Lawrence about his meeting with the two senior partners there, and in his letter of December 17, 1922, quotes himself in this dramatic exchange: "'Why in thunder didn't you secure it? It's the greatest book in the world'" (17). Although Shaw hears for the first time about Edward Garnett's abridgment, a draft of which was almost completed at this time, and would have known that this abridger was also a reader for the rival firm of Jonathan Cape, he nevertheless tries to change Lawrence's mind and have him shift it to Constables as part of an ambitious scheme: "I then expatiated on the qualities of the work, and said that it really ought to be published in the good old eighteenth century style in twelve volumes or so to begin with, the abridgement coming afterwards. This is not at all so impossible as it would have been ten years ago; for people are buying very expensive books now on an unprecedented scale, whilst my early novels in Constable's shilling series have gone out of print for the moment because cheap books do not sell" (17). Shaw mentions Garnett and says that Otto Kyllmann, one of the partners to whom he had alluded, had found out that Garnett was acting privately as an abridger and not as a reader for a publisher already selected (17). GBS pulls out all the stops in order to convince Lawrence to go with Constables. In fact, Lawrence was keenly interested in printing and publishing, almost so much so that they were as much a part of his interest in books as writing. For over a decade before Lawrence met the Shaws, he had had a particular interest in printing and the making of books. Shaw and Lawrence discussed these bookish matters at length.

Shaw's gentle and wry exasperation over Lawrence's new life in the military sums up how many, in Lawrence's lifetime and afterward, have been puzzled by his strange maneuver. As this is such a key aspect of Lawrence's life and mediates between the writer and the soldier, the hero and the man, its full rhetorical weight should be experienced whole:

> Your letter is more impressive than explicit. If I knew nothing else about you I should conclude that you were a depressed mechanic oiling up fuselages for profanely abusive pilots, and sleeping six in a bed with a hundred other such castaways. As it is, I can only pity the staff. Nelson, slightly cracked after his whack on the battle of the Nile, coming home and insisting on being placed at the tiller of a canal barge, and on being treated as nobody in particular, would have embarrassed the Navy far less. A callow and terrified Marbot, placed in command of a sardonic Napoleon after Austerlitz and Jena, would have felt much as your superior officers must in command of Lawrence the great, the mysterious, save in whom there is no majesty and might. The thing is ridiculous. Why in name of all that is sane did you not get £20,000 from parliament? It was yours for anybody else's asking, if you rule out your own; and you should have demanded it as your obvious right.
>
> You talk about leave as if it were a difficulty. Ask for three months leave and they will exclaim, with a sob of relief, 'For God's sake, take six, take twelve, take a lifetime, take anything rather than keep up this maddening

> masquerade that makes us all ridiculous'. I sympathize with them. If you must be Cincinnatus, go and farm. If you must be Garibaldi, live at Caprera instead of putting poor Aldershot out of countenance.
>
> Of course you know your own affairs worst; so I tell you how they strike me. Also, damn it, I want to know the other half of what you have told me. One can guess nothing about a man capable of anything, like Habbakuk. Are you a flying officer or a bloke in a military office? It would be so like you to be charading as an office boy. Distractedly,
>
> G. Bernard Shaw (18–19)

That Lawrence was able to answer this letter, even at times obliquely, is to his credit. While there was bluster in Shaw's flattering but penetrating satire, it would not be easy to respond to such a letter.

II

To have GBS bearing down on one's life and letters would be difficult for the most sturdy and accomplished writer. Considering what Lawrence had been through and how vulnerable he seems to have felt at the time, it is remarkable how strong and forthright he was with Shaw. It is also telling that Lawrence addresses Shaw's comments on his writing before those on his life.

In the film *Lawrence of Arabia,* the audience is witness to Lawrence's death and funeral, but not his youth, his days as an archaeologist before the war, his negotiations at the Paris Peace Talks, or his years as an enlisted man. There is no rising or falling action, only the climax of Lawrence in Arabia. There is no explanation about how he got there or what it did to him, only a suggestion that it changed him utterly and terribly, like the wound or *hamartia* that the hero sees in an *anagnorisis,* a Greek tragedy in the desert that the classically trained T. E. Lawrence, schoolboy and undergraduate at Oxford, would have recognized.

Cast a little differently, David Lean's story of Lawrence is like epic, romance and the novel translated into film. The grand narrative of Lawrence's life is not unlike Lean's other movies based on the journey of a strong central character: *Great Expectations* (1946), *Oliver Twist* (1948), *The Bridge on the River Kwai* (1957), *Dr. Zhivago* (1965), and *A Passage to India* (1984). These are stories of coming of age, education, war, revolution and empire, great sweeping tales in which heroes live, even if ambiguously, in tangled gardens, streetscapes, or exotic landscapes. Lean's Lawrence is a great adventurer and soldier who is too sensitive or vulnerable to the conflict he has so daringly mastered, a paradox of a liberator and avenger caught between freedom for the Arabs and British imperial ambitions, a great military strategist driven mad in war and by torture or rape, perhaps propelled even to war crimes, and spat out by politicians who no longer had any use for him. This is the Lawrence I knew as a child from the movie, the tall, blue-eyed O'Toole rather than the small but athletic, historical Lawrence. This TEL of history had to answer GBS, something that the film, as epic as it is, does not attempt to represent. It is hard to put Doctor Faustus in a tale about Tamberlaine, or to mix Hamlet with Henry the Fifth.

In a letter of December 27, 1922, Lawrence seizes on Shaw's phrase about *Seven Pillars of Wisdom,* that it's a "great book," which is actually Lawrence's moderating rendition of Shaw's hyperbole: "it's the greatest book in the world" (20, see 17). Lawrence, whose longstanding interest in printing is manifest in the subscribers' edition of *Seven Pillars of Wisdom,* takes up his own version of Shaw's description of the volume in terms of its layout and theme: "But you say that it's a great book. Physically, yes: in subject, yes: an outsider seeing the inside of a national movement is given an enormous subject: but is it good in treatment? I care very much for this, as it's been my ambition all my life to write something intrinsically good. I can't believe that I've done it, for it's the hardest thing in the world, and I've had such success in other lines that it's greedy to expect goodness in so technical a matter" (20). This is the core of the Lawrence I have come to know, or think I know. This Lawrence does not necessarily contradict the Lawrence of popular legend, which the film so aptly represents. But he does complicate this "Arabian prince" met in *medias res* as if the before and after barely mattered, as if the image of Lawrence could rob him of his childhood and his maturity.

Lawrence continues to Shaw, taking up the matter of writing: "However your phrase makes me hope a bit: will you let me know your honest opinion as to whether it is well done or not?" (20) The phrase is Lawrence's downsizing of Shaw's praise, even if he attributes the phrasing to GBS, as his earlier elision from "greatest" to "great" to "good" stressed the more. Humility and self-doubt are blended in this strategy of breaking down Shaw's accolade. Lawrence then proceeds with a recreation of his technique of writing and the gap between desire and performance in the creation: "When I was actually writing it I got worked up and wrote hardly: but in the between-spells the whole performance seemed miserable, and when I finished it I nearly burned the whole thing for the third time. The contrast between what I meant and felt I could do, and the truth of what my weakness had let me do was so pitiful" (20–21). What Lawrence realizes is how difficult writing is, something that does not always occur to those with lesser expectations than his. His inability to reach his ideal was a "weakness," a "flaw" that was "pitiful." I am not suggesting that Lawrence had Aristotle's *Poetics* in mind when writing this sentence, but this recognition is a catharsis without a catharsis. Lawrence is both the character in the tale and the reader following the story of a flawed hero, a hero who suffers from pity if not terror.

The *agon* of writing is something that Lawrence says that he acknowledges but avoids. To GBS, he confesses: "You see, there's that feeling at the back of my mind that if I really tried, sat down and wrung my mind out, the result would be on an altogether higher plane. I funk this extreme effort, for I half-killed myself as it was, doing the present draft: and I'd willingly dodge out of it" (21). Here is someone who was an avid student before he was a soldier, and he is returning to his pre-war aspirations to become a scholar and a writer. He who had faced so much danger now faces himself and finds the specter of writing before him. The writing of history, especially through autobiography, has hardships of its own.

Part of the unrest resides in the relation between the personal and the general in history: "Isn't it treated wrongly? I mean, shouldn't it be objective, without the first-person singular? And is there any style in my writing at all? Anything recognizably individual?" (21). The paradox is this: he desires objectivity through the expression of his own personality, through his own individual style. The autobiographical "I" rests uneasily with the third-person point of view of history.

Then there is in this letter Lawrence's move to printing and publication, which also suggests the intricate shifts of his motives, mind and character. Printing, publishing, and the decision to take a profit or not from his writings on the Arab campaign are themes that recur through Lawrence's post-war correspondence. His view on the question of writing for profit is particularly convoluted and contradictory. We have already seen him deny to Shaw any desire to make money from the Arabian writings, but here he considers it:

> About business. Curtis Brown—or rather Savage, his manager, served me in the war, and is doing my money-worries on the usual 10% terms. I hate business, and would be child's play for any publisher. I believe Cape, a new publisher of the respectable sort (he runs that divine book of extracts from yourself) is the first in the running for my thing: but I've told Savage that I want £300 a year, to live on, and have left it at that, with only two conditions, a. that I have the last word as to type, paper and format. b. that it be royalty, not an out and out sale. It's good of you to have worked up Meredith to the point of offering, and I'll tell Savage about it: but Garnett reads for Cape, and liked parts of the book: so that Cape has a special wish for it. I fancy film and serial rights are worth more than royalties: and my only motive in publishing a scrap of the book is money: so that I'm as bad as Butterworth in the matter. (21)

He disclaims any interest in or knowledge of business, but seems keenly aware of it. So often in his letters Lawrence disclaims any profit motive for *Seven Pillars of Wisdom* or any of his writings about the Arab campaign, but here he offers it as a motivation. Usually coupled with this disclaimer about profit is his justification for enlisting because the armed services will pay him a wage, which he needs in order to subsist. While Shaw was convinced that Lawrence had manufactured his own money problems by stubbornly deflecting and rejecting all honors and financial rewards, Lawrence here denies this: "So far from getting £20,000 from Parliament I had the utmost difficulty in getting my gratuity of £110 from the War Office when they mobilised me. I'm not such a figure as you think" (21). Lawrence's army file, released in 1998, seems to bear Lawrence's claim (23n5). As the letters between Lawrence and Bernard Shaw bear out, GBS himself would go on a campaign to try to get Lawrence a war honor or pension. As generous as this was, nothing came of it. The British government did not always see the Caesar in Lawrence that the author of *Caesar and Cleopatra* did.

The armed forces were not simply a means of getting away from painful memories and an intruding press and of setting up a spiritual exercise in daily routine, although this is often the case Lawrence makes in his correspondence—implicitly and explicitly. In the next part of this letter of December 17, 1922, to Bernard Shaw, Lawrence says "You ask for details of what I'm doing in the R.A.F. Today I scrubbed the kitchen out in the morning, and loafed all the afternoon, and spent the evening writing to G. B. S." (21–22). The deeper the reader gets into this paragraph, the more one sees the hero as penitent and wonders what is his relation to money and influence. The scene swings back and forth between the realm of ideas, motorbike rides, and conversations and details of menial chores: "Yesterday I washed up the dishes in the sergeants' mess in the morning (messy feeders, sergeants: plates were all butter and tomato sauce, and the washing water was cold) and rode to Oxford in the afternoon on my motor-bike, and called on Hogarth to discuss the abridgement of the Arabian book" (22). This is the prose of a skilled writer who has talent for the scene, historical, actual, and even novelistic. From butter and tomato sauce in cold washing water to his motorcycle ride to Oxford and the Ashmolean Museum, where his old mentor and friend, D. G. Hogarth, was Keeper. This is a transportation from the mess to the museum, in which the one-time colonel and war hero, Lawrence, becomes a private because of some kind of need for atonement and serves men who would have served him, and travels back, probably riding swiftly as he liked to, on a motorbike, considering his past campaigns in Arabia through a future publication with a man he had met in 1910 at about the age of twenty-two. Thus, Lawrence's journey to Oxford from the officer's mess, which he describes for GBS, was a journey backward and forward.

After describing this trip to Oxford to meet Hogarth, Lawrence then returns to his routine in the military:

> It being Christmas we do fatigues in the morning, and holiday in the afternoon. Normally I'm an "aerial photographer under training": it doesn't mean flying, but developing the officers' negatives after they land: and the 'under-training' part means that I'm a recruit, and therefore liable to all sorts of mis-employment. For three weeks I was an errand-boy, I've also been dustman, and clerk, and pig-stye-cleaner, and housemaid, and scullion, and camp-cinema-attendant. Anything does for airmen-recruits: but the life isn't so bad, when the first crudeness works off. (22)

In his letter of September 30, 1922, Lawrence had told Shaw of his enlistment and had claimed it was "a cleaner living than Eastern politics" (10). Lawrence had a good job but decided to resign and enlist; his decision was more than financial, whatever pretexts he offered. In some ways he seems to have taken pleasure in his life: "We have a bed each, and suffer all sorts of penalties unless they are 25 inches apart: twelve of us in a room. Life is very common, besides being daily. Much good humour, very little wit, but a great friendliness. They treat my past as a joke, and forgive it me lightly. The officers fight shy of me: but I behave demurely, and give no trouble" (22). Military training and monastic life are not that far apart: they are exercises of disciplining the will in order to make the body, mind, and soul fit; they are austere and communal; they can bring joy in hardship. In Lawrence's letters the reader can observe this life as something that keeps him sane and happy with its simple repetition and regulation, and that cleanses his mind and spirit by making him physically fit through wearying labor. He has to humble himself in order that the officers will leave him be. This letter, then, moves from Lawrence as writer to printer/businessman to serviceman: it brings him back to earth. Paradoxically, as that great self-publicist Shaw would have noted, this very sainthood was a means of making Lawrence into a myth. As he denied the world, it sought him out.

The British Press Association's story of December 28, 1922, about Lawrence's enlistment reads as much as an advertisement as an apologia. Concerning Lawrence, it declares: "The reason given by him for the change is that he wishes to have quiet and rest to write the concluding chapters of his book of memoirs. The opening chapters of this book, which will tell for the first time many unrecorded incidents of his adventures in Arabia and Palestine, were written at the suggestion of an Oxford friend, Captain Raymond Savage, of Messrs. Curtis and Brown, and the complete volume will be published by Mr. Jonathan Cape" (23). While this excerpt begins with what is probably a sincere motive on Lawrence's part, it shifts, in recounting the genesis of the project, to a puffing of the book. The world would not let this "saint" alone anymore than it would Saint Joan, something Shaw was probably already thinking about for his play on that soldier-saint.

III

In response to that story Shaw wrote Lawrence the same day: "The cat being now let out of the bag, presumably by Jonathan Cape with your approval, I cannot wait to finish the book before giving you my opinion, and giving it strong. IT MUST BE PUBLISHED IN ITS ENTIRETY, UNABRIDGED" (23). Shaw still hoped that Constables would publish *Seven Pillars of Wisdom* whole and then produce an abridgment if Cape would not. Shaw also introduces his wife Charlotte: "My wife, who believes fanatically in your genius, wants to know why you withdrew your preface to Doughty's book" (25). It is clear from the beginning that Charlotte helped promote Lawrence to her husband. Shaw returns to business, and offers Lawrence this warning: "The truth is I am anxious lest you should have committed yourself already. I had ten years on the managing committee of the society of authors, and learnt that there is no bottom to the folly and business incompetence of authors or to the unscrupulousness of publishers, who, being in a gambling business where one live book has to pay for ten duds, cannot afford to lose a single opportunity" (24). For whatever reason, Shaw was too late. Lawrence either did not want to make any money on *Seven Pillars of Wisdom* or had been so impractical as to ensure that he would not make anything from it.

At this time, Lawrence was thirty-four years old and the Shaws were in their mid-sixties. Charlotte Shaw's response to Lawrence is strong. Her letter of December 31, 1922, considering her age and her standing, is touching in its modesty:

> If you've been 'mad keen' to hear about your book I've been mad keen to write about it ever since I read it, or rather ever since I began to read it, and I simply haven't dared. I got from it an impression of you as an Immense Personality soaring in the blue (of the Arabian skies) far above my lowly sphere, and that anything I could say in the way of admiration, or comment, or question, could only be an impertinence. But the latest developments of your career have been so startlingly unexpected, and your letters so human, that I take my courage in both hands and send you a word. (25–26)

From the grace and aptness of these words, it becomes clear that there were two writers in the Shaw marriage. The humane nature of Charlotte (or CFS as she signs her letter) also comes out in this passage. While she praises Lawrence sincerely, she, like her husband with whom she had obviously been "communicating" concerning Lawrence and the letters he sent GBS, also wants Lawrence to settle down and realize how capable he is.

CFS actually picks up on the GBS-Lawrence exchange on *Seven Pillars of Wisdom* as a "great book." Like Lawrence himself, she is an active close reader and picks out phrases, as in pursuing an *explication de texte,* to interpret:

> How is it conceivable, imaginable that a man who could write the *Seven Pillars* can have any doubts about it? If you don't know it is "a great book" what is the use of anyone telling you so. I believe (although he has never said anything of the sort) that G. B. S. thinks you are 'pulling his leg' when you ask him. I devoured the book from cover to cover as soon as I got hold of it. I could not stop. I drove G. B. S. almost mad by insisting upon reading him special bits when he was deep in something else. I am an old woman, old enough at any rate to be your mother; I have met all sorts of men and women of the kind that are called distinguished; I have read their books and discussed them with them; but I have never read anything like this; I don't believe anything really like it has been written before. When I find in your letter such suggestions as 'should it be without the first person singular?' 'Is there any style in my writing?' 'Anything recognisably individual'? I think—are you laughing at us! Why, foolish man, it could only have been written in the first person singular: it is one of the most amazingly individual documents that has ever been written: there is no 'style' because it is above and beyond anything so silly. (26)

What more could Lawrence hope for: he had GBS and CFS, the Shaws, telling him his book was a singular wonder. CFS took up where GBS left off, saying that "Your book must be published as a whole" and "don't leave out the things an ordinary man would leave out" while ending with a recommendation that Constables would do the whole book and do it right (26–27). CFS had not liked Cape's takeover of her selections of GBS, first given to her friend A. C. Fifield, who went out of business, and so she had her own motives for pushing Constables.

But Lawrence being Lawrence, or Ross, and later Shaw, was found out, and, in a letter of January 2, 1923, tells GBS that because the *Daily Express* exposed his enlistment, he was going to change tact: "I felt that to publish anything now might look as though I were using the R.A.F. as an advertising stunt. So I've cancelled Cape's contract (fortunately not completed, and so I hope there will be no damages to pay) and have told him that nothing is to appear this year" (28). Signing himself TEL, Lawrence presents an identity that is shifting at this point. He tells Bernard Shaw, "I'm going to wash out that old name, which has too many war associations to please me: and which isn't my real name, any more than Ross!" (29).

GBS will have none of this. Taking up where CFS left off, GBS uses his considerable talent for satire to awaken TEL to his own situation and identity:

> Like all heroes, and, I must add, all idiots, you greatly exaggerate your power of moulding the universe to your personal convictions. You have just had a crushing demonstration of the utter impossibility of hiding or disguising the monster you have created. It is useless to protest that Lawrence is not your real

> name. That will not save you. You may be registered as Higg the son of Snell or Brian de Bois Guilbert or anything else; and if you had only stuck to it or else kept quiet, you might be Higg or Brian still. But you masqueraded as Lawrence and didn't keep quiet; and now Lawrence you will be to the end of your days, and thereafter to the end of what we call modern history. Lawrence might be as great a nuisance to you sometimes as G.B.S. is to me, or as Frankenstein found the man he had manufactured; but you created him, and must now put up with him as best you can. (30)

Shaw underscores Lawrence's responsibility for self-promotion and myth-making. Lawrence himself created Lawrence of Arabia and called attention to himself, so he cannot dodge his creation just because it is inconvenient or he is weary of the role. This self-conscious theatricality cannot be turned off once the drama has been produced on the public stage. Once the persona has been let out in the public domain and in history, the creator cannot control it and must put up with the good and the bad.

As Charlotte did, in a letter of January 4, 1923, Bernard also hammers at publishing the book unabridged and at "style" as something to put aside (30–31). CFS and GBS both assure Lawrence that he is above worries over style and is a writer of the highest order. Shaw faces the question of writing head on:

> As to style, what have you got to do with such dilettanti rubbish, any more than I have? You have something to say; and you say it as accurately and vividly as you can; and when you have done that you do not go fooling with your statement with the notion that if you do it over five or six times you will do it five or six times better. You set it up, and correct its inevitable slips in proof. Then you get a revise and go over your corrections to see they fit in properly and that you have not dropped one stitch in mending another. Then you pass for press; and there you are. The result has a certain melody and a certain mannerism which is your style, of which you are no more aware than you are of the taste of the water that is always in your mouth. (31)

So Lawrence finds himself debating style with one of the greatest stylists in the language who tells him there is no hothouse plant called Style. TEL is also engaged in an exchange with a person who sees writing in terms of the printing process, as he does, but from a jarringly different vantage. Shaw actually uses the word "gibe" to characterize what he has done to disabuse Lawrence of his inflated illusions (33).

GBS also has a knack for connecting the different aspects of Lawrence: war hero, writer, enlisted man, gifted publicist. He will not accept TEL's separation of these roles in time or by category. *Seven Pillars of Wisdom* provides Shaw with a way into this analysis:

> Now as to the book just as it is. You will no more be able to get rid of it, or play about it with it, than with Lawrence. It is another Frankenstein monster; and you must make up your mind to do the will of Allah, in whose hand you were only a pen. You say that to publish anything now might look as though you were using the R.A.F. as an advertising stunt. Considering that you have already used the whole Arab race and the New Testament and the entire armies of all the countries engaged in the war to advertize yourself (since you take that view of it), I do not see why you should have a sudden fit of bashfulness of the lady in reduced circumstances who cried laces in the street but hoped that nobody heard her. (31)

There is a great deal to what GBS says. Shaw points the moral—publish the book straightaway—and tries once more to land it for Constable (32). This latter aim is instrumental but it does not take away from the good sense of Shaw's argument. He is also probably right that the RAF would not necessarily mind the leak as it would advertise the force. But Lawrence has forgotten an important aspect of public spectacle that Shaw, as a dramatist, could not: "And the people have their rights too, in this matter. They want you to appear always in glory, crying, 'This is I, Lawrence, Prince of Mecca!' To live under a cloud is to defame God" (32). Lawrence, like many later celebrities, wanted the celebration of his persona on his own terms, the notoriety without being notorious. He wanted to be famous with a great deal of privacy. This is like Oscar Wilde's well-known quip that there is only one thing worse than being talked about and that is not being talked about. Apparently, Lawrence wanted to be talked about but left alone. Sometimes. The complexity of TEL's personality, as GBS himself realized, makes any definitive interpretation difficult.

That complexity makes itself known in his response to Charlotte Shaw in a letter of January 8, 1923. While Lawrence was obviously pleased by her praise, he will not yield. The letter opens:

> It's a wonderful letter, that of yours, and I've liked it beyond measure: though my doubts as to the virtues of the *Seven Pillars* remain: indeed I'd be an insufferable creature if I was sure of it, for to me a good book is the best thing that can be done. However I'd been thinking it possibly a bad book, and your praise of it makes me more hopeful. At the same time, you know, it's more a storehouse than a book—has no unity, is too discursive, dispersed, heterogeneous. I've shot into it, as a builder into his yard, all the odds and ends of ideas which came to me during those years: indeed I suspect that it's a summary of myself to February 1920, and that people who read it will know me better than I know myself. (33–34)

There is something apt in Lawrence's response. He is a writer who doubts himself, and self-doubt, when sincere, is something admirable in a writer.

He does not seem entirely satisfied with *Seven Pillars of Wisdom* and seems beyond fishing for compliments. It is also telling that Lawrence shifts from his former life in the war and its aftermath to his current life in the RAF. Whereas once he complained about the possibility of remaining an enlisted man, he now seems to have found something in it: "It would be hard to remain inhuman while jostling all those days and nights in a crowd of clean and simple men. There is something here which in my life before I've never met—had hardly dreamed of" (34). That Lawrence spent about another thirteen years in the ranks means that there was probably something to this kind of companionship and esprit that he is here describing to CFS.

TEL appears to have felt comfortable expressing his innermost thoughts to CFS. Lawrence addresses her comments on the publication of *Seven Pillars of Wisdom.* He now thinks that he is unlikely to publish the book, but that if he ever were, he agrees that "the whole is the only honest thing" (34). Very soon he returns to the personal. He has introduced Mrs. Shaw to his mother by way of her letter: "I showed my mother your letter. She likes you now, because you praised my work, and mothers have (privately) an inordinate pride in their sons. The horrors of the book strike her painfully, and she hates my having noted, or seen, such things" (34). This passage cuts to the heart of the burden of *Seven Pillars of Wisdom.* It is as much a confession as a memoir. It is Lawrence's heart of darkness: he was a great admirer of Conrad. This heavy load might explain in part his reluctance to publish and profit from the book and why he embraced the kind of quotidian routine he did in the RAF from 1922 onward. The very arguments that the Shaws had used to have him bring out *Seven Pillars of Wisdom* might have contributed to his sudden about-face.

IV

Another twist occurs in the story of Lawrence and the Shaws. On January 30, 1923, Lawrence opens his letter with an announcement to GBS that "I've now been sacked from the R.A.F., as a person with altogether too large a publicity factor for the ranks: and feel miserable about it" (35–36). At the time of this sacking, Cape agreed with Lawrence to a contract for a limited edition of 2,000 of an unabridged *Seven Pillars of Wisdom,* which, owing to TEL's state of mind, he canceled. Lawrence reiterates to GBS what he told CFS: "it's been a bad month, for the R.A.F. was the most interesting thing I ever did (after the squalid difficulty of getting used to it) and I'll regret its loss for good" (37). After having lunch with the Shaws in London on February 1, 1923, TEL announces, in a letter of March 23 of the same year, that "after hesitating here and there eventually I joined the Army, and am now at a camp in Dorset being trained" (37). In a postscript Lawrence tells GBS that he gave the Bodleian Library "at Oxford the manuscript of my Arabian effort" (38). The book would not likely be lost now.

Shaw wrote to Sydney Cockerell on April 13, 1923, that *Seven Pillars of Wisdom* "is one of the great books of our time; and I must finish it to the last morsel" (38). According to GBS's letter to TEL on May 13, 1923, Lawrence had apparently asked Cockerell to have the Shaws return the book, and Shaw makes another plea for Lawrence to get on with publishing the volume quickly and at a fair price.

On May 31, 1923, Shaw decided then to write Stanley Baldwin, the prime minister, to rectify Lawrence's complex situation. Of TEL, Shaw says:

> This man kept the Arabs fighting for us instead of with one another until the way was finally clear for Allenby to enter Damascus. Allenby sent him a message to say that he should like to have the Arabs enter with the first troops if it could be arranged. The All Souls gentleman immediately marched his Arabs into Damascus; found it left in a horrible condition of anarchy and pestilence by the retreating Turks; organized the hospitals, the police, and the municipal services in a couple of days ready for Allenby's entry; and, when Allenby arrived, asked for nothing but to be let go about his business. (39–40)

That Shaw would take this extraordinary step on Lawrence's behalf is remarkable. TEL actually went to Jesus College before the war and was a member of All Souls after, although this is not really significant to Shaw's argument. This man of many names, which GBS lists as T. E. Lawrence, Luruns Bey, Colonel Lawrence, and later Ross and Shaw, needs to be saved from himself because it is honorable for the government and for TEL. GBS's Lawrence is shy or proud or both, but he is also someone with scruples: "He was given an imaginary job in the Colonial Office; and when he threw it up as an imposture, he was left to fend for himself as best he could" (41). Nor does Shaw mince words in assessing TEL's unpredictability. Whereas Lawrence has placed the manuscript in the Bodleian and Shaw had offered him publication, "he had already realized that he had only to hold up his finger to get half a dozen offers from first rate firms. He prefers, however, to lend his five copies to selected readers, and to remain a mystery man with money and a new reputation within his reach" (41). Shaw then goes on to tell the whole story about how Lawrence deserved at least £20,000 for the services he rendered the nation but instead had trouble even collecting £130. Shaw drives his point home: "Clearly this is a bad case of Belisarius reduced to begging obols in an ungrateful country; and as the man, strange as he is, has a dramatic instinct that makes obscurity and concealment impossible for long in his case, to say nothing of this wonderful book, with its fuse burning up to a sensational literary explosion, it seems to me that the Government should consider the advisability of making a decent and indeed handsome provision for him" (42). Now TEL is one of Justinian's generals forced to beg in the streets of Constantinople. It is impressive just how highly Shaw thought of Lawrence, as "strange" as TEL sometimes seemed. Shaw dramatizes the other "Shaw": "But the fact remains that he is serving as a private soldier for his daily bread; and however much his extraordinary character may be accountable for this, it strikes all who know about it as a scandal that should be put an end to by some means. They feel that the private soldier business is a shocking tomfoolery, and are amazed to find that Lawrence is not in the position of a pensioned officer in dignified private circumstances" (43). Shaw uses a kind of royal we to add weight to the plea for a pension for Lawrence, feels "uneasy" about the situation and so wants to let the government know about it, and attests to the quality of TEL's "history" (43). By granting a pension to Lawrence, the prime minister can avoid a "scandal." Lawrence was, after all, a media magnet. Shaw refuses to see anything more than role-playing in Lawrence's enlistment.

A similar assessment of Lawrence's intricate character is made in a letter of June 4, 1923 to GBS by a person Lawrence much admired: D. G. Hogarth, to whom Shaw had sent a copy of his letter to Prime Minister Baldwin. Hogarth corrects Shaw on All Souls, which had granted Lawrence a fellowship of £300 a year from 1919 till 1926, and on the job at "the Colonial office, where his functions were more real, I think, than your phrase implies" (44). Lawrence had mentioned the magic number of £300 a year as being his desired income: he already had it and had not collected it. Hogarth continues: "The fact is that money weighs much less with him than mode of life. I cannot conceive of any government post, such as the P. M. could offer, which L. would accept, or if he accepted, retain. He begins at once to talk of 'moral prostitution' and quits!" (44). Hogarth was like a mentor to Lawrence, had known him for about thirteen years, and he was now trying to reconstruct Lawrence's motivation and way of life. In a postscript Hogarth adds: "Lawrence is not normal in many ways and it is extraordinarily difficult to do anything for him! In some measure, the life of letters is best suited to him. He will not work in any sort of harness unless it is padlocked on to him. He enlisted in order to have the padlocks rivetted on to him" (44). Here in brief is one of the paradoxes of Lawrence that puzzled many during his life and afterwards. In his next letter to Baldwin (November 12, 1923), Shaw makes use of Hogarth's information. GBS satirizes Lawrence while still asking for a pension to be offered to him or kept with the Public Trustee if, as is quite likely, TEL will refuse it. But before that, Shaw's satirical portrait is masterful: "Belisarius is still begging the obal for the sake of refusing it with dramatic gesture when it is offered. His fellowship bed and board at All Souls are still unoccupied, and his grant undrawn. He is still saluting sergeants as Private Shaw, and dashing up to London when he can get leave on his famous 80-miles-an-hour bicycle, and waits for him at Thomas Hardy's, where he is introduced solemnly as Mr. Shaw, and addressed as Colonel Lawrence" (45–46). GBS then shifts to *Seven Pillars of Wisdom* and its connections with Oxford and Cambridge:

> Hogarth of the Ashmolean and Cockerell of the Fitzwilliam are still the holders of the dead secret of his name and address, which he writes across the sky as before. Hogarth makes people (including myself) put themselves down for £10 to publish the great history of the Arabian campaign, which the author threatens to burn from time to time, having carefully presented the Bodleian with the original MS. The edition is to cost £3000; and it will be impossible for Lawrence to make anything out of it, as he insists on reproductions in colour, at great cost, of the pictures which he has commissioned from eminent artists. Cockerell declares he is fifteen different men, and that you never know which will be on duty in any given emergency. Hogarth's hair whitens when his name is mentioned. (46)

He returns to the theme with which he began this sketch and represents Lawrence as demanding the same amount he could have at All Souls:

> But all the time he wants £300 a year, and is reduced to serve as a common soldier for his bread. As far as anything can be predicted of such a man, it seems that he is determined that you shall confer this pension on him, but solely in order that he may refuse it with a wave of his panache in the manner of Cyrano de Bergerac. And always, of course, he shrinks from publicity in the middle of the limelight. In all this he is quite sincere and genuine; but he is a born actor of Protean personality, he might, for official purposes, as well as be all the Seven Humbugs of Christendom rolled into one. (46)

Shaw's TEL cannot help being an actor of many parts who has a genius for publicity even as he shuns it. What is to be done with him?

Lawrence then writes Shaw on December 13, 1923, with a scheme to have yet another edition of *Seven Pillars of Wisdom,* this a limited edition at thirty guineas for "the ungodly rich" (48). GBS replies five days later by addressing Lawrence as "Dear Luruns Bey" and by recounting what he said in the letter to the prime minister and in his interview with him. In the middle of this rhetorical flourish of reiterated satire Shaw mentions that he said "that you were a man hiding in the limelight, whispering secrets to a broadcasting loudspeaker, writing amazing Sibylline books that everybody has read, being introduced everywhere as Mr. Shaw and at once addressed as Colonel Lawrence, a man of various genius, but always an inveterate and incorrigible actor to whom ordinary reasonable action is insipid and nothing but striking gestures endurable, even a walk to the post being a meteoric rush on an 80 mph motorcycle" (49–50). GBS mentions that the prime minister will talk with Hogarth in Oxford. Shaw also warns Lawrence that the daily routine of a soldier is healthy only if he is moving on to new writing or a new campaign, "but if you are only acting, and devising superfluous ornaments for that old finished job of yours, you will presently find that you have lost the power of serious work, and that the end of pretending to be a common soldier is that you are a common soldier" (51). Lawrence replied on December 20, 1923, that the army is "an asylum for the little spirited" and says that "Your picture of my ending up to find that I am a soldier, by dint of playing much at it, comforts me: for it's the end I want, and am wanting with deadly seriousness. The peace of finding that my horizon was grown so near!" (52, 53). As Lawrence explains earlier in the letter, he wants nothing to do with the Middle East again and so would reject such a political job; nor does he wish to become an officer because he does not want to give another order (52). The question of the pension and Lawrence's motives for serving as a common soldier persisted in the correspondence between GBS and TEL. This exchange produced its own dramatic interest.

V

There is an imbalance: we have quite a few of Lawrence's letters to Charlotte Shaw but few of hers to him. It is difficult when we have his answers to her letters that are now as ghosts. TEL read the acting proof of GBS's *Saint Joan* and gave detailed comments on it. In a letter of March 16, 1924, to Charlotte, Lawrence makes a stylistic suggestion about dropping the word "pet" in one of Joan's speeches, and Shaw seems to have adopted this change (64, 67n4). He comes to the representation of Joan at the climax and finds it wanting: "I have a prejudice against the writer who leaves the reader to make his top-scene for him. . . . I funked it, in the death of Farraj, my man: faced it, in the plain narrative of my mishaps in Deraa the night I was captured" (65, see 67n6). Having read this account of that night in Deraa in chapter 87 of the 1922 edition of *Seven Pillars of Wisdom,* I can say that it is not as direct as this letter might suggest. It was difficult in the 1920s to speak openly about rape, all the more so of male rape. Even in the 1962 film, there is only a glancing sexual suggestion to the violence. So Lawrence's indirection is understandable, as is his pain in struggling to represent that event on the page. The horror at the heart of this magnificent book, this actual autobiographical event, would make the whole history all the harder to write. In another letter to Charlotte of March 26, 1924, Lawrence relates himself to Saint Joan: "The trial scene in *Joan.* Poor Joan, I was thinking of her as a person, not as a moral lesson. The pain meant more to her than the example.

You instance my night in Deraa. Well, I'm always afraid of being hurt: and to me, while I live, the force of that night will lie in the agony that broke me, and made me surrender. It's the individual view. You can't share it" (70).

However, Lawrence was able to share some of this with Charlotte, if not with GBS. The shame and pain at the heart of Lawrence's war experience was something he could not pronounce on publicly, so his behavior seemed to others odd. He sought a way to face his wound without descending into self-loathing or madness. Perhaps he gained some catharsis from confessing his feelings to Charlotte Shaw, who might understand and who would not be so close to him as his own mother:

> About that night. I shouldn't tell you, because decent men don't talk about such things. I wanted to put it plain in the book, and wrestled for days with my self-respect . . . which wouldn't, hasn't, let me. For fear of being hurt, or rather to earn five minutes respite from a pain which drove me mad, I gave away the only possession we are born into the world with—our bodily integrity. It's an unforgivable matter, an irrecoverable position: and it's that which has made me forswear decent living, and the exercise of my not-contemptible wits and talents. (70)

This wound, long before the terror of rape was discussed openly and widely, was a burden to Lawrence. How much confusion and anger and guilt arose from his experience, how it scarred him and how it affected his sexuality, is something that needs to be considered but cannot be gauged. In his confessions to Charlotte Shaw, this gloss to his own memoir, Lawrence offers that night at Deraa as the reason for his seeking out a secure life in the armed forces, a life as a private soldier doing routine tasks. This is the man who had a job at the Colonial Office and left it and who refused Air Chief Marshal Hugh Trenchard's invitation to continue the history of the air force in the war when Walter Ralegh died in 1922. Why did he not turn his back on the world straightaway? Who can say? These shocks and horrors do not necessarily work directly and in a linear fashion.

Lawrence continues: "You may call this morbid: but think of the offence, and the intensity of my brooding over it for these years. It will hang about me while I live, and afterwards if our personality survives. Consider wandering among the decent ghosts hereafter, crying 'Unclean, Unclean!'" (70–71). On sex, TEL writes to Charlotte: "Perhaps the possibility of a child relieves sometimes what otherwise must seem an unbearable humiliation to the woman:—for I presume it's unbearable. However here I'm trenching on dangerous ground, with my own aching coming to life again" (78, see 83n3). The pain of his brutal rape never left Lawrence. His letters to Charlotte are his opportunities to revisit that pain, to cleanse an old wound. Over the course of their correspondence, however, there is no steady progress, no march toward reconciliation with his past. Lawrence, although full of wit and ideas, was prone to melancholy. His various roles were not, for him, a game but an expression of anguish about a ruined past he struggled to contain. In a letter of September 28, 1925, he revisits this relation between the change in the self and the shameful wound:

> I've changed, and the Lawrence who used to go about and be friendly and familiar with that sort of people is dead. He's worse than dead. He is a stranger I once knew. From henceforward my way will lie with these fellows here, degrading myself (for in their eyes and your eyes and Winterton's eyes I see that it is a degradation) in the hope that some day I will really feel degraded, be degraded, to their level. I long for people to look down upon me and despise me, and I'm too shy to take filthy steps which would publicly shame me, and put me into their contempt. I want to dirty myself outwardly, so that my person may properly reflect the dirtiness which it conceals . . . and I shrink from dirtying the outside, while I've eaten, avidly eaten, every filthy morsel which chance threw in my way. (150–51)

Lawrence finds himself converted away from his old self through a shameful and dirty past from which he cannot recover. To Winterton, an old comrade of the Arabian campaign, and to Charlotte Shaw, his conversion seemed more of a fall, the new testament of his life a degradation of the old. The outward show of his life "conceals" the filth of the rape, humiliation, and torture. And so he hopes to find in the filth of the ranks, if not reconciliation, at least consistency.

VI

On June 8, 1926, he wrote to Charlotte, setting out the process of his composition of what would become *Seven Pillars of Wisdom:* "What were lost were my later notes: observations on the road, scribbled at random in the saddle, without much reference to place or time: they included sketch-maps, tribal notes, personal thumb-nails, complaints and things. Generally I wrote them on blocks of Army telegraph forms, which were limp, easily crushed into the fold of my belt. I have three or four of them still: (in Clouds Hill): and as many were lost" (183). Some of these notepads were lost in 1919 when the original manuscript of *Seven Pillars of Wisdom* was stolen from Lawrence at Reading Station. It is telling that Lawrence felt close enough to give Charlotte two of these notepads, which she bequeathed, along with her other Lawrence papers, to the British Library.[3] In 1926, GBS helped work on the draft preface to *Revolt in the Desert.* But GBS was growing impatient with Lawrence's perverse advertising genius, which wanted to delay the book for ten years (187–88). The final version, called a "Foreword," asks why Lawrence would abridge "an unsatisfactory book, instead of recasting it as a history," and answers, among other things, that to do otherwise he would need "an interest in the subject which was exhausted long ago with the actual experience of it" (189). Lawrence was tired of the book. But instead of publishing the whole and being rid of it, he created a limited edition of *Seven Pillars of Wisdom* that could not earn him a profit and published an abridgment.

The correspondence with Bernard and Charlotte Shaw did help Lawrence to define himself. Here he muses with the categories of action and dream. In a letter of October 5, 1926, to Charlotte, Lawrence includes a comparison between his Brough Superior motorcycle, nicknamed Boanerges, or "sons of thunder," which was capable of rest and then running "furiously," and GBS, who did everything hard without rest (68n1, 203). The order of the analogy between the motorbike and Shaw and Lawrence (like his own motorcycle) and Shaw leaves it a little uncertain as to which is given to dream and which to action: "He is like a Martha: whereas I'm like my bicycle in my capacity for sitting down, or lying down, vacantly and happily, for weeks on end. It's the difference between a man of action and a dreamer" (203). Is Lawrence the soldier actually claiming to be the dreamer while Shaw, who never rests, is the active one? If so, how is Shaw also like Martha?

Why Lawrence chose the RAF is a complicated matter, and a little like the motivation of Iago among Shakespearean critics: the ostensible motives do not always add up to the action. In a letter to CFS in November 1926, TEL claims ownership of his enlistment and absolves Winston Churchill of any responsibility:

> It is quite true that the R. A. F. is my own fault: it is my deliberate and very happy choice. I should not leave it to become Prime Minister. And it is quite true that Winston would have given me Egypt, if I had been willing, early in 1922 when Allenby first came home.
>
> (Egypt is a social job—£20,000 a year, much entertaining, a large house, a silk hat, dignity. Do you see me so?) (207–208)

Lawrence had made up his mind, at least for the moment. If he could not decide whether he had joined the RAF freely or out of financial necessity, at least he had decided that he would take no great public role, whatever the reward. Instead, he found life in the RAF to be good. In November 1926 he crashed his motorcycle and could not spend his last free Sunday with the Shaws before he went to India on a crowded troop ship (209–11). This was a hardship and an exile, as Lawrence once wrote his mother, but one that would allow him to fade from the social scene in Britain (198–99).

VII

An indirection should help provide one more gloss on TEL: the correspondence with F. N. Doubleday, his publisher in New York. Lawrence wrote to Doubleday in the 1920s and early 1930s, and his letters shed some light on his correspondence with the Shaws. Indeed there are echoes of Lawrence's words to the Shaws in these letters to Doubleday. Concerning the abridgment of *Seven Pillars of Wisdom,* he proclaims in a letter of June 30 [1920?]: "I haven't any illusions about the rotten book it is, and will feel delighted if you find it below your standard: however perhaps its faults may help it sell, and I'm a well advertised person, so that the necessary money may be raised by it. If it fails I'll have to go and work, which will be disgusting."[4] At this point, TEL does not seem to think that the service is such a good prospect. But by September 17, 1923, Lawrence is telling F. N. Doubleday to tell his wife that "Cincinnatus was probably happy, for though farming is labour, & therefore beastly, yet it is irresponsible labour: and private-soldiering though beastly & laborious, is even less responsible: & so correspondingly happier" (2 recto-verso). Cincinnatus was a Roman patrician who, as legend has it, was called from his plow to lead the Roman army to victory over the Aequi. When he had triumphed, he laid down his office and returned at once to the plow. It was to this well-known classical figure that GBS also once compared TEL.

Lawrence's letter of November 15, 1923 to Doubleday sums up many of the concerns he expressed to the Shaws about being a soldier and a writer, about his sense of self, and about his ambivalence regarding the publication of his great book.

> I'm sorry. I didn't know you had lost sight of my insignificant person. Money troubles make me hide my head. One can't be "Colonel Lawrence" (not even so mild a celebrity) on nothing a year. So I became Private Shaw, in the British Army, instead, & am paid nineteen shillings and threepence a week in compensation for the loss of my liberty & the use of my mind & body by the state. My engagement is for seven years (till about 1928) and till its termination I will not be seen in London or elsewhere. Now I am in Dorset, camped near Hardy.
>
> Naturally such a life is not my choice: I like a solitary bed, & quiet meals, & leisure for books, & the opportunity of music and converse with experienced men . . . but I had to choose between enlistment & immediate hunger, & chose to go on living. You would probably have done the same in my case . . . and I am alleviating the barrack-conditions by making myself a book-corner in the ruinous cottage for the C.C.F.L. [Cape Cod firelighters]. So that all is well (or better) now. The translating of French books pays for this luxury.
>
> I'm glad you did not close with Savage, for he was not empowered to sell my book. Indeed it is not for sale. I have been very fortunate in my readers. Kipling, Shaw, & Hardy have read it. K. did not like it much. Shaw & Hardy praised it far beyond what I ever dreamed (beyond what I still dream) its

merits. Encouraged by them I have coquetted with the idea of a reprint of its full text, to accompany reproductions of the fifty pictures illustrating it; these pictures are by some of the best artists we have & I lean greatly to see them worthily done.

Of course I don't want the book published, or even circulated. So some friends of mine are enquiring:

i. For a millionaire to put up £3000 for the privilege of owning the unique copy.
ii. For a hundred rich men to put up £30 each for a copy of a subscribed edition of a hundred copies
iii. For three hundred subscribers of ten guineas each.

The reproduction of the pictures & good printing of the text is estimated to cost exactly three thousand pounds, thus there can be no publisher and no book-sellers: and the value of a subscribed edition is that the copyright is not brought in question (either here or anywhere) & there are no reviews.

Whether any of i. ii. or iii. schemes is feasible only experiment will show. (1 verso- 2 recto)

Shaw makes an appearance in this letter in order to balance the reception of the book among great writers. And here again life in the armed forces does not seem to be a spiritual redemption, but an unfortunate necessity.

And of course Lawrence's ambivalence to publication shines through. There are a number of letters that explain Lawrence's intentions not to make any money from his books on Arabia. The way he frames the matter in these letters is very much like the way he frames the matter to the Shaws. To F. N. Doubleday on June 16, 1927, he writes from Karachi, India: "Your notion of an article from me, I'm afraid, won't do. I have decided never to use the name 'Lawrence' again: in spite of its having a market value. Nor will I write about Arabia or any derived subject. My scruples on these points have made me refuse the profits of 'Revolt in the Desert' which has sold well in England, I believe, whatever Doran may make of it and have made me keep the more valuable 'Seven Pillars' out of circulation" (2 recto). While Lawrence did translate French novels for a little money, he stuck closely to this scruple: the profits from *Revolt* went to a RAF trust. Despite his refusal to earn profit, Lawrence was pleased that Doubleday liked the art of his great tome: "Your praise for the printing of the Seven Pillars pleases me."[5]

Lawrence's own shifting identity was tied up with the fate of his book. His ambivalence for the promotion of one fed his ambivalence for the promotion of the other. In a letter of August 25, 1927, after the period of the Shaw correspondence we have examined, Lawrence discusses with Doubleday his own assumed names and pen names and explains why he can never use the surname his was born with: "So you see the name 'Lawrence' bars itself. It is worth a lot of money, because of Arabia: whereas my father chose it for me because it meant nothing, to his family. The only authentic part of my name is the initials T. E. (They do not, I believe, translate into Thomas Edward . . . but that's no matter) and most people who know me write me as Dear TE! They feel safe with that. There aren't many things safe about what are beautifully called 'natural' children!" (2 verso). Lawrence's illegitimacy now plays into his search for himself. The RAF gave him a routine, a safe haven, even an identity different from the one the world seemed to want for him. In a confessional mood, as 338171 A/C Shaw at RAF Cattewater Plymouth, he writes to F. N. Doubleday on April 4, 1929: "It does not seem to me, looking back at it, as though anything I've ever done was quite well enough done. That is an aching, unsatisfied feeling, & ends up by making me wish I hadn't done anything: and very reluctant to try anything more. Hence the RAF which is a routine that keeps me sane, & sometimes happy. A big word, that, happy: but I think it is sometimes true of me" (verso). Years before he had told the Shaws that he had found happiness in the RAF. They were still part of his life in 1929, as can be seen in this letter to Doubleday from Plymouth, dated June 28 of that same year: "Frere-Reeve wrote to me, two excellent letters, about good printing. I shall go to see him when, or if, I get leave in October. I have a design of sliding down some odd day (Dear, it will be an odd day!) if he permits it, & I achieve it, with Bernard Shaw and Mrs. Shaw. She is quaint & comfortable, and fresh, & kind. G. B. S. is exciting, per contra. Together they are like bacon and eggs, a harmony in blue & silver. I fear I talk nonsense" (1 verso).

And in his letters to Doubleday, Lawrence could not avoid self-dramatization. On December 18, 1930, he describes to Doubleday a scene with a clerk in a bookshop in Scotland: "'You are hard to please' she grumbled, offering me the Boys Book of Colonel Lawrence at a reduction, seeing I was in uniform and he now in the R.A.F. I told her I knew the fellow. And he was a wash-out: then I bought a Daily Express & escaped the shop" (3 recto). The dramatic irony here has several layers, including Lawrence's toying with his own popularization. It seems that he can see himself as a character even when others cannot see him as a person.

Sometimes there is a sense that Lawrence had stock responses for his letters. Responses from the letters to the Shaws recur in his correspondence with Doubleday. His view of himself as a writer is a case in point. While inquiring about F. N. Doubleday's failing health in a letter of January 27, 1931, he comes to the theme of writing:

> Only this writing . . . honestly, you know, I can't write. The Seven Pillars was so long ago, and it did not rise to the height of its theme. I had better stuff than any writer has had—and botched it. Then some spring or other got tired

> (there is nothing broken) and the works stood still. Today there is nothing I want to write, and I will not try to write out of an empty mind. A book should burst out of a man, against his will, like a vomit into the sea.
>
> So it is not possible for me to write: and it would be false pretences for me to accept any brain-work: there is no will for it. Body-work—yes: that I can do, and do do, all day long for the Air Force, and I do it cheerfully & well just because there is no longer any fret of the brain or ambition left.
>
> Nor was I ever a good writer, technically considered. (1 recto)

It seems that Lawrence needed more than one confessor in his life. Although he claimed that he had found happiness, for which we should take his word, TEL seems to have remained split between the writer and soldier, between the public figure and obscure everyman.

VIII

Charlotte Shaw was perceptive to see just how important a writer Lawrence was and to interest her husband in him as well. Bernard Shaw was right: Lawrence was a genius of publicity, and that genius lives on. Born late in the nineteenth century and dead a little after the first third of the twentieth century, Lawrence is still a figure in our collective imagination at the beginning of the twenty-first century. GBS saw in Lawrence's oscillation between fame and obscurity a paradoxical genius for making himself known. Shaw also realized that while TEL might not have always wanted to be recognized, he could not avoid it. Take for instance this letter to Sir John Maxwell, who had inquired after the great book:

> 4.1.26
>
> Dear Sir John Maxwell
>
> General Weight has told me that you want a copy of the private edition of my war-book. I'm willing of course—but have they told you that it won't be ready for months, that it is very long and very dull, being discreet: and that it costs thirty guineas, of which fifteen have to be sent in advance to manager Bank of Liverpool and Martins 68 Lombard St. E.C. 3. Marked "Seven Pillars Account." If you wait till March 1927 one-third of the same text will be published for sale at a guinea. Few people will want to read more, after they have read that.
>
> I've put my present name and address at the head of this letter in case you want to ask me any thing else about the book. I shall quite agree with you if you decide against it.
>
> Yours sincerely
>
> TE Shaw[6]

It is as if Alexander Pope advertised his wit by saying that he was Dullness, one of his dunces in his *Dunciad,* and somehow convinced people that he was the wittier for it. Maxwell would not be put off by all this talk of tediousness or by this undermining business maneuver, making a cheap edition available in order to undercut the subscribers' edition (not something Pope would have done in the eighteenth century when he set up a subscription for his Homer). Neither Maxwell nor the world would ignore Lawrence, despite his best, albeit ambivalent, efforts.

But that is Lawrence, man of the book. And then there is Lawrence, man of the film. I recently watched the film for the first time since its first incarnation in 1962. In the archives at Princeton University I found a book entitled *Columbia Pictures Presents the Sam Spiegel and David Lean Production of* Lawrence of Arabia.[7] Through its title page Peter O'Toole's eyes look out through a rectangular cutout from his full picture on the next page. His face is framed by the lines "His name will live in" and "English letters" (A1 recto-A2 recto). In the unpaginated film book, A2 recto proclaims: "'I deem him one of the greatest beings alive in our time. I do not see his like elsewhere. His name will live in English letters; it will live in the annals of war; it will live in the legends of Arabia.'—Winston Churchill." And GBS is twice invoked as part of this puffing of Lawrence: "'With the single exception of myself, no man of our time has had such a power of tempting journalists and even diplomats to tell lies about him as Lawrence.'—G. B. Shaw." About *Seven Pillars of Wisdom* the book declares: "But George Bernard Shaw gave him enthusiastic encouragement in its preparation, H. G. Wells called it a great human document, E. M. Forster judged it a masterpiece but feared to tell Lawrence so lest he evoke a sarcastic reaction from a man whose friendship he treasured. Certainly it is unique."

His correspondence with the Shaws, who do not appear in the film, is one way to understand Lawrence as a writer who wanted to escape the past that he had created, a man who had to face a wound he would rather not. He could not turn to fiction in the face of a truth that was his heart of darkness. That darkness remains. The film makes a strong statement when it shows Lawrence meeting his death on the motorcycle he so loved. Danger was one of his idols, this man of shifting identities. The death at high speed in the prime of life is something to remember. But the student and scholar before the war and the serviceman afterwards, hard as Lawrence tried, could not be ordinary. Like the saint who has confessed and the sinner who has converted, like the soldier as court jester and holy man, Lawrence was more than of Arabia.

Notes

1. T. E. Lawrence, Jeremy and Nicole Wilson, eds., *Correspondence with Bernard and Charlotte Shaw 1922 - 1926* (Fordingbridge: Castle Hill Press, 2000); unless otherwise noted, page numbers in this essay refer to this edition.

2. See Jeremy Wilson, *Lawrence of Arabia: The Authorised Biography of T.E. Lawrence* (London : Heinemann, 1989), 637–38.
3. Add. MS 45914 and 45915; see Wilson and Wilson, eds., 184n2.
4. Frank N. and Nelson Doubleday Collection, Box 7, Manuscripts Division, Department of Rare Books and Special Collections, Princeton University Library.
5. T. E. Lawrence to F. N. Doubleday on June 16, 1927, 2 verso, ibid.
6. T. E. Lawrence to Sir John Maxwell on January 4, 1926, in Sir John Maxwell Papers (1859–1929), B Correspondence 1 General I-M Co 583, AM 19367, Box 4, Manuscripts Division, Department of Rare Books and Special Collections, Princeton University Library.
7. New York: Richard Davis, 1962.

Chapter Seven

T. E. Lawrence and the Art of War in the Twenty-First Century

George W. Gawrych[1]

If there were a hall of fame for modern military theorists, Thomas Edward Lawrence would deserve a place in it. In his dual role of theorist and practitioner of the art of war, Lawrence demonstrated the power of military theory for developing appropriate strategy and tactics in war. In working effectively with the leaders of the Arab Revolt in World War I, he left insights for forging a successful coalition to defeat a common adversary. By embracing the Bedouin Arab way of war, he was able to develop a theory of guerrilla warfare that still holds relevance today. In light of the above achievements, Lawrence should stand as a model for military officers as they prepare intellectually and emotionally to face the challenges of the twenty-first century.

Lawrence's exact role in the Arab Revolt, it must be noted, remains a subject of much controversy. Lawrence certainly has had his many detractors and skeptics.[2] There is no denying, however, that Lawrence is worthy of commendation for his service in Arabia. Not only did he see extended combat in one of the harsher environments of the world, the Arabian desert, but he clearly played an important role in a number of significant military operations. For assisting the Arabs in capturing the port of Akaba in July 1917, for example, Lawrence received a promotion to the rank of major from the British Army and the *Croix de Guerre avec palme et citation a l'ordre de l'Armée* from the French government. The Ottomans, for their part, offered a five thousand pound reward on his head.

Overall, Lawrence did help coordinate Arab military operations with the British war effort against the Ottoman army in Palestine. Yet his exact achievements will always be subject for disagreement. But as Jeremy Wilson, the leading British authority on Lawrence, has noted: "After the war, fellow-officers who had seen his work at first-hand said that his contributions had been outstanding. Although some of these witnesses may have exaggerated, others were men of high integrity. Their testimony cannot be entirely groundless."[3] Whatever his exact role in the military operations, Lawrence proved a keen observer and a gifted writer. This is the legacy that concerns us here.

Background

By the time of his arrival in the Hejaz in October 1916, T. E. Lawrence was ready to carve for himself a place in modern military history. At the age of fifteen or sixteen, Lawrence began reading books on military history and theory. He entered Jesus College at Oxford University in 1907 with a view of studying modern history. His passion for military subjects led him in 1909 to spend four months in Syria and Lebanon conducting research on Crusader castles. After returning to England, he completed his bachelor thesis in 1910, entitled *The Influence of the Crusades on European Military Architecture—to the End of the XIIth Century,* which was eventually published as *Crusader Castles.*[4] From 1910 to 1913, Lawrence conducted archaeological research in the Middle East at Carchemish on the Upper Euphrates. With war in Europe on the horizon, the British Army enlisted his service in helping map the Sinai Peninsula, a project that resulted in the publication of a book entitled *The Wilderness of Zin* (1915).[5] The fieldwork, conducted under the supervision of a regular British officer, exposed Lawrence to a systematic evaluation of the military value of terrain.

The outbreak of World War I found Lawrence fully engaged in London completing his study of the Sinai. Upon its completion, Lawrence joined the army as a lieutenant. Because of his experience in the Middle East, within a short time the British Army assigned him to its intelligence branch in Cairo. There he renewed his acquaintance with Gertrude Bell (1868–1926), whom he first met in May 1911. This Englishwoman came to Cairo to work in the British intelligence service because of her extensive experience living and traveling in the Middle East. She proved an invaluable source for firsthand knowledge of Arab tribes and tribal chiefs, no doubt providing Lawrence with useful insights into Bedouin society for his future assignment in Arabia.[6] Her information complemented that which Lawrence had gained from his own readings on Arab history and society, including the famous work by Charles Montague Doughty (1843–1926), *Travels in Arabia Deserta,* originally published in 1888. Doughty has been regarded as the greatest of all English travelers in Arabia, and he certainly commanded Lawrence's great admiration.

Lawrence's assignment to the intelligence branch in Cairo proved an excellent final step in preparation for work among the Arabs. Here, for almost two years, Lawrence was able to amass "an encyclopaedic knowledge of the Ottoman Empire, and also of the Turkish army and its dispositions. Each day an immense amount of military and political information passed through his hands."[7] This steady flow of information attuned Lawrence to the higher issues of warfare, the interplay of policy and military operations, especially beginning in August 1915 when Sherif Hussein, the emir of the Islamic holy cities of Mecca and Medina, first offered the British an Arab alliance against the Ottomans.

By October 1916, Lawrence was well prepared for his assignment to Arabia. As a result of serious study, extensive travel, and several years of residence in the Middle East, Lawrence was proficient in the Arabic language. He also possessed an impressive knowledge of Arab society, a knowledge that served him well in Arabia as a British liaison and advisor to the Arabs in revolt against Ottoman rule. His intelligence work gave him a good understanding of the strategic background for the British-Arab alliance. As a citizen-soldier, he found it easier to transcend the general conservatism of professional officers and to coordinate Arab military operations with the British campaign in Palestine and Syria. Well-read in military literature, Lawrence possessed the intellectual sophistication necessary to articulate his military experiences and observations into a coherent theory of irregular warfare. All this stands as no mean accomplishment for even a regular officer.

Lawrence and Military Theory

When Lawrence landed in Jidda on October 16, 1916, the Arab Revolt was in full swing. Sherif Hussein, the emir of the Islamic holy cities of Mecca and Medina, had revolted, with British and French assistance, against Ottoman rule in Arabia. Arab tribes loyal to Hussein had already captured most of the Hejaz, including Mecca but not Medina. Hussein commanded his tribal armies through his four sons Abdullah, Feisal, Ali, and Zeid. Initially, British command in Cairo sent Lawrence to Arabia in order to gather intelligence on the revolt. Rather quickly, however, he gained the confidence of Sherif Feisal and remained in the Hejaz to serve as his British liaison.

In advising the Arabs, Lawrence came to appreciate the importance of military theory, the power of placing under one yoke intellect and action. Basil Liddell Hart, his most famous military biographer, claimed that "As a boy T. E. always thought that he was going to do great things, both 'active and reflective'—'I hadn't learned you can't do both'—and determined to achieve both."[8] Certainly by the time he reached Arabia, Lawrence realized that successful action demanded critical thought. Effective thinking, in turn, required an appropriate theory or philosophy of war. There in Arabia during World War I, Lawrence turned to military literature for understanding and guidance. His successful service in the Arab Revolt attests to the importance of theory for sound action in war.

Advising the Arabs demanded a sophisticated understanding of war. As a military advisor, Lawrence, now a captain, faced more than a tactical problem. Politics and strategy weighed more heavily than tactics in such an assignment.[9] Lawrence had to link the Bedouin style of warfare with the Allied goal of defeating the Ottoman army in Palestine and Syria. This task required him to strike a harmony between discordant interests and strategies, a formidable undertaking even for a regular officer. Yet Lawrence proved up to the task.

Fortunately, Lawrence possessed the intellectual background necessary for this work. His studies in modern history included a solid foundation in both military history and military theory. "In military theory, I was tolerably read," Lawrence once remarked.[10] This statement is quite modest, but his professional reading was impressive by the standards of any day. Indeed, he was familiar with the works of Clausewitz, Jomini, de Saxe, Moltke, du Picq, Guibert, von der Goltz, and Foch.[11] These theorists were all noted for their insights into conventional warfare.

Sound military theory is essential for conducting war at a level higher than the tactical. Without it, officers must rely only on their own training, experiences, and intuition. Sound military theory, on the other hand, develops the intellect. It provides an intellectual framework for analyzing the essence of war. It allows officers to analyze war in its various forms, such as conventional war, guerrilla warfare, civil war, or people's war. It develops their critical thinking and judgment; officers learn how to think, rather than what to think. This is exactly what Lawrence needed in his role of liaison to the Arab Revolt: a theoretical framework from which to conduct a systematic analysis of his theater of operations.

Initially, Lawrence had little time to reflect seriously and critically about his theater of operations. Upon arriving in the Hejaz, Lawrence saw "a crying need for action,"[12] so he uncritically relied too much on instinct in developing his initial courses of action. However, in March 1917, a combination of boils, dysentery, and malaria laid him up in a tent in Abdullah's camp in Wadi Ais for some ten days. During this rather lengthy convalescence, Lawrence turned to serious thought and critical analysis. He searched, in his own words, "for the equation between my book-reading and my movements."[13] He sought to connect theory with practice, the abstract with the concrete. Only then could a compass be found with which to negotiate through the Bedouin world of warfare.

Maurice de Saxe (1696–1750) provided Lawrence the theoretical foundation for his own theory of guerrilla war. Writing nearly two centuries earlier, in 1732, de Saxe found that generals were too preoccupied with tactics, marches, and formations and that they therefore ignored the intellectual aspects of war: "very few men occupy themselves with the higher problems of war. They pass their lives drilling troops and believe that this is the only branch of the military act."[14] As a result of this preoccupation, these generals were neglecting critical analysis of "the higher problems of war," including those of strategy and campaign planning. Like Lawrence, de Saxe had difficulty finding time to dwell on these higher problems. And like Lawrence, de Saxe had to suffer illness to gain the time necessary for serious reflection and writing. During thirteen days of convalescence, de Saxe wrote *My Reveries Upon the Acts of War,* a treatise on war that remains a classic today.[15] This work provided Lawrence with a theoretical framework for appreciating of the Bedouin way of war in Arabia.

Lawrence left his own body of military literature, perhaps modest in quantity but certainly impressive in quality. His *Seven Pillars of Wisdom* stands out among all his writings. Unlike most military memoirs, which are mainly a record of personal challenges and triumphs, *Seven Pillars of Wisdom* can be read as an insightful study of Arab tribesmen and their way of war. Lawrence also left for future generations of officers a significant article entitled "The Evolution of a Revolt," published in 1919.[16] In only fifteen pages, Lawrence succinctly articulates his theory of guerrilla warfare, one that also appears in expanded form in *Seven Pillars of Wisdom.* In addition, in 1917, while in the midst of his assignment in Arabia, Lawrence filed "Twenty-Seven Articles," a report in which he offered principles for working effectively with Arab Bedouins in coalition warfare.[17] Together, these three writings constitute excellent professional reading for the intellectual development of officers, especially in the art of irregular and coalition warfare.

The United States Army should place a premium on military theory in the education of its officer corps. From a safe distance, it appears to do so. "Know thyself and know thy enemy" goes the famous dictum in the profession of arms. The military academy at West Point, the staff college at Fort Leavenworth, and the war college in Carlisle Barracks stand as august educational institutions with global reputations. The staff and war colleges are meant to prepare officers for the higher levels of war, and military theory should have a prominent place in the curriculum of both schools. A close examination of the United States Army, however, reveals a very different reality. American military culture enshrines technological superiority as the United States Army's magic bullet in warfare. Officers generally shun military theory, and the army values field experience over schooling in promotion of its officer corps. "Those who can't [command], teach" has been a popular attitude in the army. This anti-intellectual statement suggests a dichotomy between thought and action, a dichotomy that the best officers have always sought to overcome.

The Vietnam War stands as a tragic example of this anti-intellectualism in the armed forces. Professor Douglas Pike, a noted expert of that conflict, evaluated the attitude of the United States military toward the Communist Vietnamese as "vincible ignorance": one does not know, one realizes one does not know, and yet one feels no need to change the fact that one does not know. Pike summarized the lack of critical thought thus: "No high-level permanent institution was created to analyze enemy strategic thinking. . . . No significant biographical studies of enemy leaders were done. . . . One can search the voluminous *Pentagon Papers* in vain for extended discussions of the other side, any discussion at all. . . . Work on order of battle generally was good; politics of the Politburo was hardly touched."[18] General Bruce Palmer, Vice Chief of Staff of the United States Army from 1968 to 1973, concurred with Pike's assessment: "Yet, the truth is that the US was basically ignorant about the enemy's *character and strategy*" (my italics).[19] According to this line of reasoning, the enemy was all tactics and no strategy, and American technology and idealism would defeat his national will. No hard, critical, reflective thinking was necessary. Rather, the quest for maintaining the initiative and seeking a decisive battle remained the holy grail of United States military operations. Superior technology would provide the means for victory.

Despite the tragedy of the Vietnam War, anti-intellectualism remains a part of American military culture today. The profession of arms in the United States is generally not a reflective one. In an article published in 1984, Huba Wass de Czege, then a colonel in the United States Army, underscored that American military culture encourages its officers to be doers rather than thinkers, disposing them to seek practical solutions to problems rather than rely on theoretical approaches to problem-solving. Practical knowledge and personal experience far outweigh the importance of military theory, and the latter is often viewed as too abstract to have any concrete value. In direct response to this general attitude in the United States Army, Wass de Czege, who eventually rose to the rank of brigadier general, argued that military theory should serve as a foundation for meeting the challenges of reform and innovation.[20]

Little has changed in the United States Army since the publication of Wass de Czege's article some sixteen years ago. Personally, with each passing year at the United States Army Command and General Staff College (at Fort Leavenworth, Kansas), I am more and more surprised at the general lack of serious reflection or analysis by individual officers or by the larger military institution on how the United States Armed Forces, together with its allies, defeated the Iraqi army in the Gulf War.[21] In fact, today, the United States Army is too busy integrating new technologies into its force structure in preparation for its next major war to encourage critical examination of how the Gulf War experience might unconsciously be shaping that preparation.[22] Critical introspection suffers as a result of a heavy reliance on technology for solving military challenges in the future.

Lawrence's service in Arabia, however, stands in sharp contrast to the general bias against military theory in American military culture. Lawrence demonstrated the importance of linking theory to practice. His insights into the nature of guerrilla and coalition warfare warrant serious consideration.

Military Theory of Irregular Warfare

To help him fulfill his mission in Arabia, Lawrence had to appreciate the Arab way of war. As underscored by Clausewitz nearly a century earlier, "The first, the supreme, the most far-reaching act of judgment that the statesman and commander have to make is to establish . . . the kind of war on which they are embarking, neither making it for, not trying to turn it into, something that is alien to its nature. This is the first of all strategic questions and the most comprehensive."[23] Lawrence fulfilled this need admirably.

Warfare falls into two broad categories: conventional and unconventional. Conventional warfare refers to wars fought between regular armies in traditional battles, with the primary military objective being either to defeat the adversary's army or to seize and hold strategic terrain. To ensure victory in conventional war, commanders are expected to bring large forces together in mass and firepower in order to defeat an opponent's army. Battle becomes the centerpiece of strategy and tactics. Unconventional warfare, on the other hand, refers to irregular warfare, people's wars, or guerrilla warfare. This kind of war requires a very different mindset and very different tactics and strategy. Irregular forces lack the men and material to engage a regular army in a major battle, and therefore they resort to a strategy of defeating small parts of the main army, most often with hit and run tactics. Bedouin Arabs were centuries-old practitioners of guerrilla warfare.

To achieve success in Arabia, Lawrence needed to understand the nature of guerrilla warfare. Here, he came to embrace de Saxe as his military mentor. Lawrence regarded the eighteenth-century Austrian general and theorist as "the greatest master of this kind of war." De Saxe offered a theory of war based on the model of a general who practiced the dictum that "a war might be won without fighting battles."[24] Whether Lawrence was aware of this or not, others had presented a similar ideal. Some 2,500 years earlier, Sun Tzu, the most famous Chinese theorist of war, wrote that "the highest excellence is to subdue the enemy's army without fighting at all."[25] By arguing for a military strategy based on maneuver without battle, de Saxe provided Lawrence with a theoretical framework for negotiating the Arab world of war with purpose and direction.

Lawrence could see direct application of de Saxe's theory of avoidance of battle to the unconventional warfare of Arabia in the twentieth century. In Bedouin society, battle carried an importance markedly different from that of European mass armies waging war on the continent. The Bedouin generally nurtured a sensitivity to high casualties. Europe's bloody battles of World War I waged over weeks or months made no sense in an environment where the population was scarce, the desert vast, and the organization tribal. In Arabia, seizing fertile ground, maintaining personal honor, or capturing prize booty often carried more weight in developing military strategy than the conventional aim of defeating an army in bloody engagements.

De Saxe's theory of war without battle as a centerpiece of military strategy gave Lawrence a theoretical base from which to analyze and appreciate the Bedouin way of war. Inspired by de Saxe, Lawrence developed his own concept of a "war of detachment."[26] Avoid seeking the enemy's strength in battle; instead, conduct a strategy based on raids by 100 to 200 tribesmen against targets designed to unbalance the adversary. "Our tactics were always tip and run, not pushes, but strokes."[27] To deny the Turks lucrative targets, the Arabs naturally resorted to the principle of employing "the smallest force, in the quickest time, at the farthest place."[28] In other words, the Arab strength lay in employing a raiding strategy. Battle should be engaged only under the most favorable conditions. This was the essence of Bedouin warfare.

The main elements in Lawrence's theory of irregular warfare may seem obvious and simple today. In his day, Lawrence experienced an institutional bias in the British Army toward conventional strategy and tactics: "We all looked only to the regulars to win the war. We were obsessed by the dictum of Foch that the ethic of modern war is to seek for the enemy's army, his center of power, and destroy it in battle."[29] Although Sherif Hussein wanted European weapons and technicians to help him defeat the Ottoman army at Medina in a major battle, the Bedouin Arabs couldn't be transformed into a Western army. They rejected formal discipline and the training programs designed to break individuality for the purpose of forming cohesive combat units. Instead, the tribes preferred to fight under their own *sheikhs* as individual warriors and as members of tribes. Even an adept Arab leader such as Feisal could not easily mix tribes together. And when Arab regular units were imported from Egypt, Syria, or Iraq, tensions between them and the tribes often hampered military operations.

Lawrence understood these limitations and was frustrated by them at times. In August 1917, after the capture of Akaba by the Arabs, he wrote in a letter to Clayton: "Of course it would be nice and much simpler for us if the Arab Movement emerged from the bluff-and-mountain pass stage, and become a calculable military problem: but it hasn't yet, and isn't likely to."[30] Despite all the Arab military weaknesses, Lawrence still found much to be admired in the Bedouin way of war: "Do not try to trade on what you know of fighting. The Hejaz confounds ordinary tactics. Learn the Bedu principles of war as thoroughly and as quickly as you can, for till you know them your advice will be no good to the Sherif. Unnumbered generations of tribal raids have taught them more about some parts of the business than we will ever know."[31] Where others focused on Arab weaknesses, Lawrence saw, in the words of General Palmer, "character and strategy" among the Bedouins.

In developing his own theory of irregular warfare, Lawrence identified three key elements for analysis: the algebraic, the biological, and the psychological.[32] In understanding the interplay of these three elements, he was able to appreciate the strategy and tactics that would allow the Arabs to play a complementary role in the British effort to defeat the Ottoman army in Palestine and Syria. A careful reading of his short analysis helps officers sharpen their own analytical skills.

"The algebraic element of things" refers to the physical environment that has shaped warfare in the Hejaz. For Lawrence, this was the decisive element. Here war is part science, depending on mathematical calculations with which to analyze the fixed condition of time, space, and terrain. Using simple math, Lawrence calculated the size of territory held by the Arabs in relation to the number of Ottoman troops in theater. The Ottomans, with only 16,000 troops in Arabia and with a shortage of staunch Arab allies among the tribes, lacked enough soldiers in order to establish effective control over 140,000 square miles of territory. Geography, the vast desert, gave the Arab Revolt sanctuaries that the Ottomans could not seize and hold for any length of time. As noted by Lawrence, "to make war upon rebellion is messy and slow, like eating soup with a knife."[33] The Arabs possessed safe havens in the vast desert and therefore had the time to conduct a protracted struggle. They received critical assistance from the British Army, an army that posed a serious threat to Palestine. Consequently, the Ottoman High Command felt it could ill afford to spare additional troops to quell the Arab uprising. Foreign assistance and a distracted enemy proved a window of opportunity for the rebellious Arabs.

Lawrence was right in his systematic analysis. Safe havens, foreign assistance, and an undermanned opponent are important factors in helping an insurrection achieve success. The Vietnamese communists used safe havens in Laos, Cambodia, and North Vietnam to defeat the United States; the Afghans, for their part, relied on Pakistan and the mountainous terrain of Afghanistan to place a nail in the Soviet coffin. In each case, one superpower provided vital assistance to a smaller ally to help defeat the rival superpower. Moreover, both the United States and the Soviet Union limited their war efforts. Each, for example, set limits on its troop levels so as to not overstrain the home front. Neither superpower, therefore, committed enough troops to secure the countryside. By placing major constraints on themselves, both superpowers played to the enemy strengths based on geography, time, and will. Meanwhile, both the Vietnamese and the Afghans were committed to total war and thus were willing to engage in bloody battles.

After analyzing geography, Lawrence next addressed the human dimensions of warfare, which he called "the biological element of lives." Here, war is part art, for human beings are involved in waging it. Intangibles such as genius, fear, heroism, and morale lay outside the domain of quantitative analysis. The irrational exerts its own powerful influence over military operations. Biologically speaking, Arab tribesmen were masters of the raid, capable of employing strategic mobility across vast stretches of desert. However, unlike the Vietnamese or the Afghans, the Bedouins were disinclined to wage bloody battles with heavy casualties in a total war effort.

Yet the art of war includes both the human and the material. The Ottoman Empire was beset with economic woes so that, according to Lawrence, the loss of material proved a greater drain on resources than the loss of soldiers. The Arabs could turn the Ottoman material weakness into their own strength. Without a heavy reliance on a base of operations for logistics, Bedouin warriors could easily disappear into the vast desert, only to appear suddenly elsewhere to destroy a bridge, cut the railway, seize a supply train, or overrun an outpost. In a bolder move, the Arabs could, through a strategic maneuver, suddenly attack and defeat an Ottoman garrison. Such was the case when, much to the surprise of the British, slightly over five hundred Arabs seized Akaba on July 6, 1917, after having traversed inhospitable desert to attack the port city from an unexpected direction. The Arabs could move with stealth through the desert, appearing at the appropriate time for an attack. The result was, in the words of Lawrence, "a vapor, blowing where we listed."[34] Strategic mobility was an Arab strength, offsetting their weakness in sustaining casualties. "Our cards were speed and time, not hitting power, and these gave us strategical rather than tactical strength. Range is more to strategy than force."[35]

But against what specific target or targets should these raids focus? In Cairo, Lawrence had spent much of his time analyzing the Ottoman army in Arabia. Initially, he viewed Medina as the locus of Ottoman military power whose conquest would give Sherif Hussein a decisive victory. But a well-entrenched Ottoman garrison defended the second holiest city in Islam, and the Arabs lacked the conventional power and will to seize this prize at a price of heavy human loss. After the Ottomans failed to take Rabegh en route to capture Mecca, Lawrence and the Arabs captured Wejh, a small town some two hundred miles north of Rabegh. From here, the rebellious Arabs posed a direct threat to the Hejaz Railway, the only communication and supply link between the Ottoman garrison at Medina and

Palestine. Some five hundred miles of railway separated Medina from Ma'an in Transjordan. The Hejaz Railway thus served as the lifeline of the Ottoman army in Arabia, a lifeline vulnerable to attack.

In order to protect his only line of communication, the Ottoman commander had to divide his force in two, one to protect Medina and the second to guard the railway. By this decision, he effectively lost a maneuver force to challenge the Arab tribes with offensive operations. The desert thus became an even greater sanctuary for the Arabs as the strategic initiative now clearly passed to Sherif Hussein. As Lawrence noted at this juncture of the war effort, "perhaps the virtue of irregulars lay in depth, not in face."[36] The mere threat to the railway, backed by sporadic raids, was enough to pin down some sixteen thousand Ottoman troops in Arabia. The Ottomans saw their military power immobilized as they placed their main effort on defending Medina and the railway.[37] The Ottoman garrison at Medina held out until the end of the war, left to wither on the vine of the Hejaz Railway.

Finally, the third element in Lawrence's military theory was "the psychological element of ideas." Initially, Lawrence failed to grasp this dimension: "I had not seen that the preaching was victory and the fighting a delusion . . . as Feisal fortunately liked changing men's minds rather than breaking railways, the preaching went better."[38] Here was the imperative of gaining and maintaining legitimacy for a rebellion through spreading the word. The use of force or threats to convince tribes could lead only to internecine tribal warfare. Propaganda, whenever appropriate, was a more effective tool. According to Lawrence, psychological warfare had to target three main audiences: one's own troops, those of the enemy, and the civilian population, in this case the townspeople and the tribes. Their guerrilla warfare had to be presented as a struggle based on a noble cause: Arab independence from Turkish rule.[39] Propaganda helped forward the Arab Revolt; the printing press served as a useful instrument in this regard. Today, CNN and the Internet have replaced the printing press, but the nature of the problem is much the same. In a rebellion, ideas are important in the quest for legitimacy and loyalty.

In irregular warfare, the importance of national will, or, in this case, tribal opinion, is hard to exaggerate. Lawrence thus came to understand and appreciate the importance of this psychological element. After all, war is a social phenomenon, and irregular warfare takes on the dimension of a struggle for the hearts and minds of the people. Lawrence, however, was under no illusions. He understood that the Bedouin tribes were often won over to the cause not so much by words as by the money provided by the British to the tribal chiefs.[40] Primary loyalty remained with the tribe, and its collective action could often be bought for a price.

In developing his theory of irregular warfare, Lawrence embraced the Bedouin world unlike his peers. The tribes taught him much about warfare in the desert. He came to understand the limited nature of his conflict and avoided trying to transform the Arab Revolt into something alien to its nature. He learned to appreciate the interaction of the material, the human, and ideas in such warfare. He gave proper attention to the factors of safe havens, foreign assistance, a dispersed enemy, and a friendly population. In the end, Lawrence proved quite successful in articulating the essential features of guerrilla warfare.

Coalition Warfare

True to his desire to marry action and contemplation, Lawrence gave serious thought to his mission within a British-Arab coalition. He came to admire the Bedouin warriors and recognized their character and strategy. Many British officers, however, failed to share his admiration for the Bedouin and instead affected a superior attitude and behavior. Others were willing to learn, but needed instruction. To address this general problem, Lawrence felt compelled to offer advice on bridging the cultural gap between the British world and that of Arabia. In his "Twenty-Seven Articles," he offered practical wisdom for those British officers assigned to Arabia.[41] He published this report a month and a half after the Arabs had captured Akaba. The underlining message was quite clear: the Bedouins were worthy of admiration for their unique way of war. Despite particular circumstances, many of his principles for dealing with the Bedouin serve as excellent advice for officers assigned to work in any coalition.

In his report, Lawrence underscored the necessity of openness and flexibility. His introduction cautioned that "handling Hejaz Arabs is an art not a science, with exceptions and no obvious rules."[42] One could make the same statement for any coalition partner. There are no easy answers or shortcuts to gaining an understanding of a foreign society. To be effective in Bedouin society involved acquiring as much information as possible about the region's leaders and the tribes themselves. Near the end of the report, he emphasized that "the beginning and ending of handling Arabs is unremitting study of them. . . . Your success will be just proportional to the amount of mental effort you devote to it."[43] In conducting his own intelligence assessments, Lawrence stressed the importance of learning power relationships among the Bedouins. British officers, for example, had to appreciate the difference between a *sherif* and a *sheikh* in Bedouin society.[44] And Lawrence remained true to his own advice. As late as June 1918, he was hard at work gathering and analyzing material on the political loyalties and history of the northern tribes as the Arab Revolt moved north from Akaba to Damascus.[45]

Military commanders generally do not seek such knowledge, though publicly all would champion its value. Over eighty years later, for example,

General Norman Schwarzkopf, the commander of Western forces in Operation Desert Storm, would fall short on this score. Khaled bin Sultan, who commanded the Arab coalition in the Gulf War, wrote of the American general: "the people, the leading personalities of Arab politics, the families, the customs, attitudes, language, history, religion, way of life—indeed all the complexities of our Arab world—were as foreign and unfamiliar to him as they are to the average American."[46] As commander of United States Central Command responsible for the Middle East, Schwarzkopf should have possessed some depth of knowledge on these subjects, at least enough to impress his Arab hosts. But in this regard, he failed.

In addition to cultural knowledge, Lawrence argued for a proper attitude toward one's ally in order to avoid unnecessary friction and problems. He counseled patience, respect, tact, and even a good dose of humility. The foreign officer had to take time to ingratiate himself into the inner circle of a tribe in order to gain its trust. He had to resist the temptation to give orders or to seek the spotlight at the expense of his hosts. Tribes would naturally resist following foreigners, and it took effort and skill to have them coordinate their military action with that of a Christian nation. To help maintain the coalition, Lawrence advised the sharing of glories with an ally, if at all possible: "Strengthen his prestige at your expense before others if you can."[47] He championed common courtesy in dealings with the Arab: "If we are tactful, we can at once retain his goodwill and carry out our job."[48]

Writing years later, Khaled bin Sultan unknowingly confirmed much of Lawrence's advice. The Saudi general found some fault with Schwarzkopf's attitude and behavior in dealing with Arabs during Desert Shield and Desert Storm: "I believe he never fully grasped my overriding concern to ensure that we did nothing during the war that might compromise out postwar future."[49] For instance, there could be no hint that Saudi sovereignty was in any way compromised in deference to American power. As Khaled noted, "My public appearance as the Saudi commander had to be as impressive as his, down to the smallest detail."[50] Meetings between the two commanders, for example, had to take place in Khaled's office. Schwarzkopf could not have more bodyguards or vehicles in his entourage than those possessed by his Saudi counterpart.[51] Such seemingly little things mattered much to the Saudis. They were deeply concerned about legitimacy of the Saudi regime, which already had been compromised, to some degree, by the king inviting Western troops into his country. Had Schwarzkopf read Lawrence's "Twenty-Seven Articles," he would have better prepared for dealing with Saudi leaders.

No doubt, many of Lawrence's insights into the nature of coalition warfare derived from direct observation. Feisal was attempting to organize a coalition of Arab tribes that failed to see themselves as a single nation. Tribal chiefs guarded their independence fervently. In reality, Feisal lacked unity of command. Rather, he commanded by the consent of the Bedouin tribal chiefs. In such a fragile coalition of Arab tribesmen, Feisal had to be more of a diplomat than a commander. He had to be careful not to alienate tribal leaders with orders but rather to coordinate military operations through verbal persuasion, often laced with monetary and material incentives. He was, in fact, attempting to lay the foundations for a future state as well as waging a war against the Ottomans. Maintaining this all-important coalition proved his main effort.

Arab tribesmen expected Feisal to play the traditional role of tribal *sheikh*. In this regard, Lawrence merely depicted his patron as that ideal. According to Lawrence, Feisal gave access to all, never cut short petitions, showed extreme patience and self-control, demonstrated goodwill and humor, and exhibited tact by never allowing anyone to leave his presence "dissatisfied or hurt."[52] Cynics and skeptics haved criticized Lawrence as a Western imperialist who portrayed Feisal as a "noble savage." Yet Lawrence's description conformed to the tribal leader idealized by the Bedouins themselves. Moreover, this style of leadership makes sense in coalition warfare. Dwight D. Eisenhower practiced it to some degree as supreme Allied commander in Europe during World War II. He, for example, demanded collegiality and courtesy from subordinates and staff, tried to reach decisions by consensus, and devoted most of his time to coalition politics. To avoid unnecessary friction, American soldiers received a booklet instructing them on British customs and habits.[53] For demanding deference to the British, Eisenhower received criticism from fellow Americans for catering, in some instances, to the British at the expense of the Americans.

Lawrence approached his assignment with a mindset similar to that of Eisenhower in World War II. He also stressed flexibility, adaptability, and collegiality in dealing with the Bedouin. His guidelines, however brief and focused on Bedouin society, remain a valuable source for addressing proper attitude and behavior in any coalition. National arrogance and cultural insensitivity remain sources of friction and antagonism in any multinational war or peace support operation. American officers, as they study the nature of coalition warfare, should give serious thought to Lawrence's insights on the subject.

Study of Lawrence's military career in Arabia does point to one major challenge in coalition warfare: competing interests among allies. On this score, Lawrence found himself serving two masters. He represented British interests to the Bedouins but also attempted to champion the cause of Arab independence, especially as the rebellion moved north toward Damascus and, after the war, in the peace negotiations in Paris. Lawrence felt duplicitous for his part in negotiating the labyrinth where these two worlds—British and Arab—intersected politically and militarily. The effort left Lawrence with guilt. He came to believe that he had failed the Arabs.[54]

American officers might learn from Lawrence that they must be prepared to strike a harmony between competing interests in multinational operations. The stakes are always high, and they are often personal as well as political.

Conclusion

Much to his credit, Lawrence demonstrated the importance of military theory in warfare. Officers today should study his career and writings for the ways in which they challenge the profession of arms.

First, officers should set aside time for serious reflection on their profession, and military theory is an important tool in this endeavor. Reading the military classics is essential for intellectual development in peacetime. Second, coalitions, whether in war or peace support operations, will remain a feature of the United States' military deployments in the future. Lawrence offered some necessary principles and guidelines that are indispensable for officers who are assigned to work with an allied army.

Third, officers should study Lawrence's theory of guerrilla warfare and the way he arrived at it. As the only superpower left after the Cold War, the United States Army continues to focus its main effort on preparing for conventional warfare. Currently, however, peace support operations absorb much of the army's time and resources, and the military establishment rather reluctantly embarks on these missions. Meanwhile, army education devotes little attention to unconventional warfare. In preparing for the challenges of the twenty-first century, the United States Army must educate its officers in the entire spectrum of conflict, from global war to peacekeeping. It can ill afford to be selective in its preparations. For his part, Lawrence offers a military theory that effectively addresses guerrilla warfare, that part of spectrum between conventional wars and peace support operations.

Regardless of the controversies about his personal life, Lawrence stands a model of the importance of wedding theory and practice in warfare. The United States Army relies too heavily on technological superiority to compensate for the anti-intellectualism of its military culture. Lawrence's military career in World War I attests to the importance of intellectual preparation of officers. The twenty-first century promises to have many surprises across the spectrum of conflict. Studying Lawrence will help prepare American officers for the unexpected challenges of the profession of arms in the future.

Notes

1. The views expressed here are those of the author and do not reflect in any way the position of the Combat Studies Institute, United States Army Command and General Staff College, Fort Leavenworth.

2. See, for example, Suleiman Mousa, *T. E. Lawrence: An Arab View* (London: Oxford University Press, 1966).
3. Jeremy Wilson, *Lawrence of Arabia: The Authorized Biography of T. E. Lawrence* (London: Heinemann, 1989), 2. This works remains the best and most comprehensive study of Lawrence.
4. T. E. Lawrence, *Crusader Castles* (Oxford: Clarendon Press, 1988). The thesis was first published in 1936.
5. C. L. Wooley and T. E. Lawrence, *The Wilderness of Zin* (London: Jonathan Cape, 1936).
6. Wilson, *Lawrence of Arabia,* 232.
7. Ibid., 197.
8. Quoted in Ibid., 21.
9. Tactics refers to the art and method of fighting individual battles or raids, whereas strategy is the art of linking of linking battles to an overall concept and method of winning the war and achieving the political aim.
10. T. E. Lawrence, "The Evolution of a Revolt," *Army Quarterly* 1 (October 1920 and January 1921): 57; T. E. Lawrence, *Seven Pillars of Wisdom* (New York: Penguin Modern Classics, 1962), 193. This paper does not address how much Lawrence read of these theorists or how well he understood each one. Even today, few military historians or professional soldiers have read all of Clausewitz's *On War,* but many readily quote him. Lawrence based his ideas much more on personal experience than on critical and serious study of military theory.
11. Lawrence, "Evolution," 57–58; Lawrence, *Seven Pillars,* 193.
12. Lawrence, *Seven Pillars,* 193.
13. Ibid.
14. Maurice de Saxe, "My Reveries Upon the Acts of War," in *Roots of Strategy,* edited by Thomas R. Phillips (Harrisburg, PA: Stackpole Books, 1985), 296–97.
15. Ibid., 300.
16. T. E. Lawrence, "The Evolution of a Revolt," *Army Quarterly* 1 (October 1920 and January 1921): 55–69.
17. T. E. Lawrence, "Twenty-Seven Articles," *The Arab Bulletin: Bulletin of the Arab Bureau in Cairo, 1916–1919,* Volume II, 1917 (London: Archive Editions, 1986), 347–53. For a recent reprint see Wilson, *Lawrence of Arabia,* 960–65.
18. Douglas Pike, "Conduct of the War: Strategic Factors, 1965–1968," in *The Second Indochina War: Proceedings of a Symposium held at Airlie, Virginia 7–9 November 1984,* edited by John Schlight (Washington, DC: Center of Military History, 1986), 113–14.
19. Bruce Palmer, "Commentary," in Ibid., 153.
20. Huba Wass de Czege, "How to Change an Army," *Military Review* 64 (November 1984): 32–49.
21. Currently, the staff college offers a ten-month course to some 1,000 officers. Some 780 come from the army and 120 from other services. In addition, approximately 90 international officers attend the course as well.

22. I base this evaluation, in large measure, on my own observations from teaching at the staff college since 1984. Naturally, I have also discussed this subject with numerous faculty members, both military and civilian, as well as with students over the years.
23. Carl von Clausewitz, *On War,* edited and translated by Michael Howard and Peter Paret (Princeton: Princeton University Press, 1984), 88–89.
24. Lawrence, "Evolution," 68.
25. Sun Tzu, *Sun-Tzu: The Art of Warfare,* translated by Roger T. Ames (New York: Ballantine Books, 1993), 111.
26. Lawrence, "Evolution," 61.
27. Ibid., 64.
28. Ibid.
29. Ibid., 65.
30. Wilson, *Lawrence of Arabia,* 439.
31. Lawrence, "Twenty-Seven Articles," 351.
32. Lawrence, *Seven Pillars,* 197–202 and "Evolution," 59–62. For a succinct analysis of his military thought, see Lawrence Moores, *T. E. Lawrence: Theorist and Campaign Planner* (Fort Leavenworth, School of Advanced Military Studies, 1992): 11–38. Another good source is Basil Liddell Hart, *T. E. Lawrence in Arabia and After* (London: Jonathan Cape, 1934).
33. Lawrence, "Evolution," 60.
34. Ibid., 59.
35. Ibid., 63.
36. Ibid., 57.
37. Ibid., 57, 58, 61
38. Lawrence, *Seven Pillars,* 178.
39. Ibid., 180.
40. Michael Asher, *Lawrence: The Uncommon King of Arabia* (New York: Overlook Press, 1999), 262–63.
41. Lawrence, "Twenty-Seven Articles," 347–53. A full copy of this report also appears in Wilson, *Lawrence of Arabia,* 960–65.
42. Lawrence, "Twenty-Seven Articles," 347.
43. Ibid., 352–53.
44. Ibid., 351.
45. Wilson, *Lawrence of Arabia,* 514–16.
46. Khaled bin Sultan, *Desert Warrior: A Personal View of the Gulf War by the Joint Forces Commander* (New York: Harper Perennial, 1995), 208. This book is certainly not without the author's own biases, including some stereotyping of Americans.
47. Lawrence, "Twenty-Seven Articles," 348.
48. Ibid.
49. Khaled bin Sultan, *Desert Warrior,* 208.
50. Ibid., 204.
51. Ibid., 192.

52. Lawrence, *Seven Pillars,* 125–28. Lawrence also wrote several uncomplimentary remarks about Feisal's character and leadership. Feisal was not a perfect leader; neither was Lawrence an unbiased critic. Friction certainly existed between the two men, and both used each other for political ends.
53. David Eisenhower, *Eisenhower: At War, 1943–1945* (New York: Random House, 1986), 158–61.
54. Lawrence is quite candid about this dilemma in the opening pages of his book. See *Seven Pillars,* 21–24.

Index